Early Praise for *Build a Weather Station with Elixir and Nerves*

I came to this book without any Nerves experience, but left feeling empowered. The authors demonstrate how powerful Nerves is and how we can build useful hardware projects with very little code. I highly recommend this book, especially if it's your first hardware project.

➤ **Stephen Bussey**
    Founder, Clove and author of *Real-Time Phoenix*

The project is very clear without being too simple. Looks really fun and I know from experience that Nerves is a great way to get some hands-on experience with Elixir.

➤ **Lars Wikman**
    Founder and CEO, Underjord AB

A great introduction to Nerves with a practical project that also succeeds as an introduction to powerful tools such as Docker Compose, Grafana, and TimescaleDB.

➤ **Anderson Cook**
    Software Engineer, dscout

A compelling interleaving of systems design and embedded development knowledge that uncovers a productive and approachable workflow with Nerves.

➤ **Jason Johnson**
    Co-owner, FullSteam Labs

Build a Weather Station with Elixir and Nerves

Visualize Your Sensor Data with Phoenix and Grafana

Alexander Koutmos

Bruce A. Tate

Frank Hunleth

The Pragmatic Bookshelf

Raleigh, North Carolina

Many of the designations used by manufacturers and sellers to distinguish their products are claimed as trademarks. Where those designations appear in this book, and The Pragmatic Programmers, LLC was aware of a trademark claim, the designations have been printed in initial capital letters or in all capitals. The Pragmatic Starter Kit, The Pragmatic Programmer, Pragmatic Programming, Pragmatic Bookshelf, PragProg and the linking *g* device are trademarks of The Pragmatic Programmers, LLC.

Every precaution was taken in the preparation of this book. However, the publisher assumes no responsibility for errors or omissions, or for damages that may result from the use of information (including program listings) contained herein.

For our complete catalog of hands-on, practical, and Pragmatic content for software developers, please visit *https://pragprog.com*.

The team that produced this book includes:

CEO: Dave Rankin
COO: Janet Furlow
Managing Editor: Tammy Coron
Development Editor: Jacquelyn Carter
Copy Editor: L. Sakhi MacMillan
Layout: Gilson Graphics
Founders: Andy Hunt and Dave Thomas

For sales, volume licensing, and support, please contact *support@pragprog.com*.

For international rights, please contact *rights@pragprog.com*.

ISBN-13: 978-1-68050-902-1
Book version: P1.0—January 2022

Contents

Acknowledgments

It has most likely been said before, but we think it is worth saying again: writing and publishing a book that you are genuinely proud of takes a village. What you see before you would not have been possible without the people mentioned in this section, and we would like to take the opportunity to thank them for all of their time and effort.

We would like to start by thanking the staff at The Pragmatic Bookshelf for all of their help and support. Working with The Pragmatic Bookshelf has been both a pleasure and an honor. A special thanks must also be given to Jackie Carter, our editor, who worked tirelessly to ensure that the book we produced was all that it could possibly be. For a couple of us, this book was our first publication, and Jackie was always there to guide us when we needed help.

Creating a project-based book that is both concise and a pleasure to read was no easy task, and a huge thanks is needed for our technical reviewers who helped us realize this goal. Steve Bussey, Anderson Cook, Sophie DeBenedetto, Jason Johnson, Nikos Maroulis, Parker Selbert, and Lars Wikman all provided the critical feedback necessary to ensure that the final book was technically correct, easy to follow, and flowed naturally.

While also being an author of this book, a big thank you should also be extended to Frank Hunleth for all of his hard work around Nerves. Without all of his contributions to the Nerves framework and the ecosystem of tools surrounding Nerves, this book would not have been possible. Similarly, we would also like to thank Justin Schneck for all of his contributions to Nerves and the supporting ecosystem, as the tools would not be what they are today if it were not for his efforts.

A special thanks is also in order for the creator of the Elixir programming language, José Valim. The Elixir programming language and community has made a profound impact on all of our careers, and we are all extraordinarily grateful for everything that José does to push the language, community, and ecosystem forward.

Lastly, we would like to thank our families and friends for all of the love and support while we were working on this book. Creating this book was a labor of love, and it would not have been possible if it were not for the amazing people that surround us.

Alexander Koutmos

I would like to personally thank my wife, Carol, and our two daughters for all of the love, happiness, and joy that you give me every single day. The three of you are the source of my strength and motivation that allow me to create things such as this book, and I am truly blessed to have you all in my life.

Bruce Tate

Writing is a labor of love, and my heart goes out to those who make it so. Maggie, you are my joy and inspiration always. You stood by me with joy as the tiny blinking LED grew to a full solder station, a bookcase, and more spare parts than Tony Stark ever had. I love you!

Frank Hunleth

I would like to thank my coauthors for including me in this project. Being able to share the enjoyment I get from working with hardware with both of you was a blast. And Christy, thank you for making it possible to follow my passions and bringing so much happiness to our family.

Introduction

The Erlang virtual machine and the Elixir programming language have managed to find their way into a wide range of software domains. We believe that this is very much in part due to the technical merits of the BEAM runtime and the phenomenal developer experience that the Elixir programming language provides. With both of these elements at your disposal, you truly feel like you have superpowers when programming with Elixir.

This feeling of having "superpowers" is immediately apparent if you've ever written, maintained, or deployed a Phoenix application.[1] The fault tolerance, scalability, and ease of development that you get from the Phoenix Framework and Elixir is a thing of beauty. Not to mention that if your requirements include some sort of real-time user interaction, the tooling available to you in the Elixir ecosystem is second to none.

While Elixir has enjoyed great success in the API back end and web-application space (thanks to Phoenix LiveView[2]), what if you could realize these same "superpowers" in an embedded systems and IoT context? That is exactly what the Nerves Project aims to do.[3] With Nerves, you can develop reliable and fault-tolerant IoT applications without compromising productivity. Nerves takes care of all of the lower-level concerns, leaving you with only the application-specific concerns. In addition, since you have the power of the Erlang virtual machine at your fingertips, you can create concurrent real-time applications leveraging all of the abstractions that you already use.

As you'll see in this short book, Nerves allows you to create very capable IoT applications in record time, without the pain and frustration of building everything from scratch. In less than 100 pages you'll have an end-to-end solution for capturing and visualizing weather data, from the embedded Nerves application capturing and publishing data to the Phoenix application that

---

1. https://phoenixframework.org/
2. https://hex.pm/packages/phoenix_live_view
3. https://www.nerves-project.org/

receives and persists the data. You can have a full stack Elixir solution all the way from the embedded-hardware level without even breaking a sweat!

What You Will Build

To get hands-on experience with Nerves, you'll be building a simple yet powerful weather station. This book starts at the hardware layer first and focuses on getting started with Nerves. After you get your Nerves-powered device up and running, you'll start writing the code necessary to retrieve sensor data from your weather station sensors. After you get your Nerves application code to a point where it needs to publish its collected data, you'll shift your focus to a lightweight Phoenix JSON API. The Phoenix JSON API will store the captured weather data into TimescaleDB for efficient time-series querying.[4] Once the Phoenix API is up and running, you'll switch back to the Nerves application code and write a lightweight HTTP-based data publisher that will push environmental metrics to your Phoenix back end. With all the weather data eventually finding its way into TimescaleDB, you'll then leverage Grafana to visualize all of the time-series data.[5]

How to Read This Book

This books takes you step by step through the process of building an end-to-end IoT weather station—from data collection, data persistence, data visualization, and everything in-between. As such, it's strongly recommended that you read this book cover to cover, as omitting steps may result in a nonfunctional end product.

Who This Book Is For

This book is for any Elixir programmer that is comfortable with the basics of the programming language and is interested in dabbling in the world of embedded systems. No soldering or deep hardware experience is necessary, given that you'll be working with off-the-shelf plugin-and-play hardware.

Who This Book Isn't For

If you're just starting off with Elixir (welcome, by the way!) or struggle with concepts such as GenServers and pattern matching, we highly suggest picking up a more beginner-oriented Elixir book prior to starting this one. Luckily there are some great resources out there such as Programming Elixir 1.6.[6]

---

4. https://www.timescale.com/
5. https://grafana.com/
6. https://pragprog.com/titles/elixir16/programming-elixir-1-6/

While Elixir 1.6 came out a few years ago now, the core language hasn't changed much in that time, so the book will help you develop a solid Elixir foundation. After you read that book, feel free to pick this one up again and get your hands dirty with an IoT-based project.

Running the Code Exercises

Being able to build and run your application code will be key to understanding the concepts outlined in this book. As such, it's important that you have the items outlined in the next couple sections so that you have everything you need to complete the weather station project.

Software Requirements

Embedded hardware aside, you'll need the following:

- Elixir version 1.9 or greater
- A Linux, MacOS, or Windows machine to do your development on
- A Linux, MacOS, or Windows machine to run your Phoenix API Server
- A wireless access point for your local area network
- The ability to run Docker containers

If you have all of those items, then you are good to go from a development machine perspective, and all that's needed is the Nerves-related hardware.

Hardware Requirements

While there is some flexibility with what hardware (like what version Raspberry Pi) you can buy and from where, the following items were used by the authors:

- Raspberry Pi Zero W with headers
- Qwiic pHAT v2.0 for Raspberry Pi
- VEML6030 Light Sensor (Qwiic connection)
- BME680 Environmental Sensor (Qwiic connection)
- SGP30 Air Quality Sensor (Qwiic connection)
- Qwiic Connection Cables
- MicroUSB connection cables
- 4GB+ MicroSD card
- MicroSD card reader

If you don't know what these things are or where to buy them, fear not—we explain all of this in the first chapter.

Online Resources

All of the code for this project can be found online at the Pragmatic Programmers web page for this book[7] or in this GitHub repository.[8] If you need any assistance for all things Elixir and Nerves, be sure to check out the Elixir Forums where you'll find a vibrant community ready to help.[9]

---

7. https://pragprog.com/titles/passweather
8. https://github.com/akoutmos/nerves_weather_station
9. https://elixirforum.com/

CHAPTER 1

Elixir and Nerves for IoT

In this book, we'll be exploring the capabilities of the Elixir programming language paired with the Nerves IoT (Internet of things) platform. To understand how real-world Nerves applications are developed and structured, we'll be building a network-enabled IoT weather station. The Nerves-powered weather station will be able to collect a wide array of sensor data and will then publish that data to a lightweight Phoenix back-end application which will be backed by a time-series database for efficient data storage and retrieval.

As you'll see through the various stages of the project, having the power of the Linux kernel at your disposal allows you to iterate quickly and to focus on the important parts of your project—getting your weather station to read sensor data and publish it over the network. Pair that with the power of the concurrent, functional, and pragmatic Elixir programming language and you have a recipe for a powerful yet simple, network-enabled, embedded IoT device.

You'll start off the weather station sensor hub application by creating a vanilla Nerves project and pushing that firmware to a Raspberry Pi. After you've burned an initial firmware to the device, you'll work on getting it onto the wireless network so that you can communicate with your IoT device, even when it's not hooked up to your development machine. From there, you'll learn about organizing your Nerves projects and reading sensor data over I2C. With all of your sensors hooked up and fetching data, you'll add a component to your Nerves application to publish data over the network.

Once you start developing network-enabled applications, the main features of Elixir and OTP quickly jump to the forefront. Since networked software must often wait for responses from remote servers, concurrency becomes vitally important. In addition, given that complex software is more likely to crash, the ability for OTP to restart itself becomes crucial. Finally, dealing with complexity is easier in a higher-level language with more and better

Frank says:
How Does Nerves Relate to Linux?

It's important to differentiate the Linux kernel from all the libraries, shells, utilities, and other software found on Linux systems. Nerves uses the Linux kernel for basic operating services and device drivers such as those needed to use WiFi or operate a camera. Nerves configures the Linux kernel so that it starts the BEAM and your applications on boot.

Since OTP provides such a rich set of services out of the box, the usual Linux utilities aren't needed, and Nerves provides slim and efficient base images by default. In the cases where you can't avoid a Linux program or service, though, Nerves provides a way to include those.[a] Elixir and the BEAM are always front and center with Nerves, but you're not locked into only using those technologies.

---
a. https://github.com/nerves-project/nerves/blob/main/docs/Customizing%20Systems.md

abstractions; Elixir's functional programming model also presents plenty of opportunities for higher productivity.

Before diving into the weather station project though, it's important to first discuss what IoT is at a high level and why Nerves is a good fit for the IoT space.

Why Nerves for IoT?

IoT is all about having interconnected computational devices geographically distributed to where they are able to perform their required work. In the automotive industry, for example, an IoT device may be used to track the whereabouts of a vehicle via GPS, detect when heavy-braking events occur using accelerometers, or even capture data from the engine's ECU (electronic control unit) to see how it's operating with regards to fuel economy. All of these "things" can be measured and captured by IoT devices and then sent via WiFi, Bluetooth, LTE, or NFC (near-field communication) to other systems for further processing or long-term storage.

This need to capture and relay data from dynamic environments, is something that spans many industries and sectors. IoT devices are being used in sectors such as manufacturing, supply chain and logistics, agriculture, smart homes, and even healthcare. What Nerves brings to the table when it comes to developing IoT applications is the ability to use consumer-available hardware (Raspberry Pi, BeagleBone, and even x86 powered hardware) without having to build everything from scratch.

Specifically, Nerves provides the tooling and foundation to get a device from zero to running a vanilla application in less than thirty minutes. A bunch of community libraries are available to help you configure sensors, without having to write any low-level code. Having the ability to deploy robust and fault-tolerant IoT devices powered by Elixir, Nerves, and OTP can be a strategic advantage for your business, or just plain fun if this is something that powers a hobby of yours.

As you'll see throughout this book, Elixir (and by extension Nerves) can provide you the tools you need to get your IoT project up and running in record time!

Time-Series Sensor Hub

To experience the robustness of Elixir in an IoT setting, you'll be leveraging the Nerves framework and its many tools to build a sensor hub weather station. The IoT sensor hub will collect weather data at a regular interval and then publish that data to a Phoenix RESTful API. So that you can retrieve this weather data for later review, your Phoenix server will be persisting that sensor weather data into PostgreSQL.

While we could use a vanilla install of PostgreSQL for this project, the nature of our data is time-series, and it would be best if we leverage a time-series database. Specifically, it will be far more performant if we queried our sensor data from a persistent datastore that supports time-series data as a first class citizen. Luckily for us, a PostgreSQL extension solves this exact problem and it's called TimescaleDB.

By leveraging the TimescaleDB extension, you get all the benefits of using PostgreSQL as well as utilities for dealing specifically with time-series data. Under the hood, the TimescaleDB extension will automatically partition your data by time and allow you to interact with this data as if it were all contained within one database table. The database table that you interact with is also known as a hypertable and is merely a facade for all of the time-sliced partitions.[1] This takes all the administrative overhead out of manually partitioning tables and creating new partitions as days/weeks roll over in the database. From the Elixir side of things, given that we are still interacting with a PostgreSQL database, all of our database interactions still take place using Ecto.

As you'll see from working on the project throughout this book, the pairing of a time-series database with an IoT sensor hub is a very powerful technology stack. For this particular project, we'll be connecting to a wireless LAN and

---

1. https://docs.timescale.com/latest/introduction/architecture

Alex says:
Partitioning by Space

In addition to partitioning your data efficiently by time, TimescaleDB also allows you to partition your data by space. What this means is that you can tell TimescaleDB to further partition each hypertable by an additional dimension to extract even more performance out of TimescaleDB. This additional dimension is derived from your table's schema and can be any other column aside from the timestamp column that you used to create the hypertable.

Leveraging space partitioning can be useful when you have a dataset that is well suited for time and space partitioning, but it can also decrease performance if the dataset isn't well suited.[a] In the context of IoT applications, it may be useful to space partition inbound data by the device that it was published from or by a geographical region where multiple devices have been deployed to.

---

a. https://docs.timescale.com/timescaledb/latest/how-to-guides/hypertables/best-practices/#space-partitions

will be publishing the measurements to a server running on the LAN. While in a real-world setting these IoT devices may find themselves in remote locations well out of reach of a WiFi network, the proposed setup is suitable for this application.

If you're interested in building Nerves projects that run in remote environments, be sure to check out the VintageNetMobile[2] and VintageNetQMI[3] projects to see how you can leverage cellular modems from your embedded devices.

Laying Out the Architecture

Given that there are a number of components to this project (a time-series database, a nerves sensor hub, and a Phoenix back-end API) and they all operate at different layers, it would be beneficial to first visualize all the parts and how they interact with one another. Let's take a look at the architecture diagram on page 5 and break down how the various components will work with one another.

At the bottom left of the diagram you'll notice that we have entries for Nerves Weather Stations 1, 2, and N, which denotes that we can arbitrarily scale our IoT fleet up and down as the need arises. In a real-world application, you may have a vast number of sensor hubs deployed, all of which are reporting back

---

2. https://github.com/nerves-networking/vintage_net_mobile
3. https://github.com/nerves-networking/vintage_net_qmi

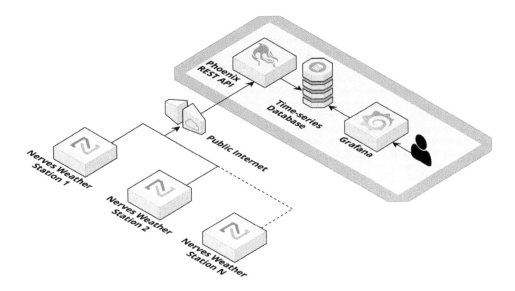

to your server-side application. Thus the Phoenix API acts as the gateway for all of the sensor data collected by your IoT devices.

Between our fleet of IoT sensor hubs and our Phoenix API is a network interface. For the weather station application that we'll be building throughout this book, that network interface is the WiFi antenna built into the Raspberry Pi, and our home LAN. For production use, this network interface could be ethernet or LTE,[4] depending on where your IoT device is deployed to.

Once our data is pushed from our Nerves IoT devices to our Phoenix API (via HTTP), our server-side application will persist that data into our TimescaleDB-enabled PostgreSQL instance. One of the reasons that TimescaleDB is a good fit for the problem at hand is that it deals well with high-cardinality data.[5]

What Is Cardinality?

Cardinality, as it pertains to the data stored in a database, is a measure of how many different values are present for a particular field or column. For example, if you're using UUIDs to capture a user's id, you will have high cardinality because each user will have a unique ID. In other words, if you have 50,000 users, you will have 50,000 possible values for the id column.

---

4. https://github.com/nerves-networking/vintage_net_mobile
5. https://blog.timescale.com/blog/what-is-high-cardinality-how-do-time-series-databases-influxdb-timescaledb-compare

> **What Is Cardinality?**
>
> On the other hand, if you have have a column called user_type, for example, and that is an enum with possible values of basic, admin, and super_user, you would have low cardinality for the user_type column. The reason for this is that even if you have 50,000 users stored in your database table, user_type can only be one of three possible values.

After our Phoenix application has persisted our data into PostgreSQL, its responsibilities with regards to that data are effectively over. We could possibly extend our Phoenix application to return this data via a RESTful API or even present the time-series data via a LiveView SVG chart.[6] But in the spirit of getting things up and running quickly, we'll lean on the powerful yet simple data visualization tool Grafana.[7]

Grafana has the ability to connect to a wide array of data sources like Prometheus,[8] InfluxDB,[9] and even PostgreSQL+TimescaleDB.[10] This will allow us to host our own instance of Grafana (via Docker), which we can then interact with, to visualize all of our time-series data as it is persisted into our database.

With a high-level understanding of how all the pieces fit together and how they communicate with one another, it's time to dive deeper into the Nerves side of things and see how you'll be structuring your Nerves project.

Organizing Your Nerves Project

Before writing any code, it's important to first understand the features that will be required for the weather station and how we can go about organizing the project structure. As you'll shortly see, Nerves projects have their own special organization structure that makes it really convenient for embedded Elixir applications.

Planning the Features

Fundamentally, the weather station sensor hub project is composed of sensors, a REST API interface, and a firmware project to manage the application life

---

6. https://github.com/mindok/contex
7. https://grafana.com/
8. https://prometheus.io
9. https://www.influxdata.com
10. https://grafana.com/docs/grafana/latest/datasources/postgres/

cycle. Let's be a bit more specific, though. Here are the individual features we need to build:

- Each sensor will need to have its own GenServer and will be responsible for collecting its sensor's measurements.

- Our network-enabled Raspberry Pi will have a GenServer that periodically captures measurements from the connected sensors and publishes them to our Phoenix REST API.

- The firmware project will be responsible for building the Nerves firmware, starting all the sensor GenServers, and also starting the HTTP publisher GenServer. This will act as the glue that brings together all our dependencies and our externally sourced dependencies.

Let's dig a little deeper into how we'll be structuring the code for the weather station.

Organizing the Project

If you've worked with Elixir projects before, you may know that they come in two general flavors. The first being regular stand-alone Elixir applications and the other being umbrella applications. The former can be used to house a single Elixir application where dependencies are pulled from external sources (like Hex, for example). The latter, on the other hand, allows you to have multiple applications all colocated within the same mono-repo. In addition, you can also have dependencies in the umbrella project for use by the separate applications. While you can organize your Nerves applications in either of the aforementioned ways, there's another recommended method for organizing Nerves projects called the *poncho project*.

In short, poncho projects provide an opportunity to isolate the Nerves-specific code from dependencies, protect major interfaces, relegate complexity to one layer at a time, and separate the concerns of packaging and distribution. Poncho projects achieve this goal by grouping separate Elixir projects within the same top-level directory structure. If an Elixir application in the poncho project needs to reference a dependency that is also in the poncho project, it does so by leveraging the path: "../adjacent_elixir_app" keyword list entry in your mix.exs file.

This is particularly useful since each application in the poncho project can maintain its own configuration files and you can burn your firmware from the poncho application that is specific for Nerves. This is in contrast to an umbrella project where all the applications share the same configuration files.

Frank says:
How Poncho Projects Came to Be

Early on, many people were interested in using Phoenix on their Nerves devices to provide local management interfaces. Both Phoenix and Nerves have enough complexity on their own that combining them in the same project was difficult. At the time, technical issues prevented Elixir's umbrella project from working, and Justin Schneck found that using simple path dependencies in a mono-repo worked well and was easy to explain. He coined the term *poncho project* for one of our early training classes to differentiate the strategy from the more well-known umbrella project.

With that being said, let's see how we'll lay out the project. Here's the poncho project structure that we'll follow (we'll start building this in the next chapter, but it's useful to see how the project will be laid out):

```
└── sensor_hub_poncho
    ├── publisher
    ├── sensor_hub
    └── veml6030
```

The sensor_hub_poncho top-level folder will typically be the one that serves as your repository. It's not an Elixir project per se, as you'll create it with a plain old mkdir instead of a mix new or mix nerves.new. Each of the subprojects within sensor_hub_poncho is then a regular Elixir application with its own config/ directories and mix.exs files.

Important to note here is that the sensor_hub subproject is your *firmware project*. Firmware projects only deal with configuration, life-cycle concerns, dependencies, and glue code. As a result, you'll need to add your other subprojects to sensor_hub as you build out additional functionality and project dependencies.

Ponchos also often have two types of dependency projects. Hardware dependency projects don't need firmware because they usually write to hardware interfaces. These projects are usually completely independent of specific configuration. The veml6030 project is a great example of a *hardware dependency* project, as it will only wrap the ambient light sensor.

The second kind of dependency project—the hardware-independent dependency—is more interesting. If you think about it, placing as much code as possible in hardware-independent layers makes sense because development on hosts is often more productive than working on targets, since there's no firmware update cycle. Whenever you can, it's important to move code from hardware dependencies into hardware-independent dependencies. Our publisher subproject will provide our HTTP API client implementation and is a great example of an independent project. You'll build the publisher subproject so that it doesn't need to integrate with specific hardware features.

Firmware projects will connect to dependency projects using path dependencies.[11] For the example, the sensor_hub subproject will have path dependencies to the publisher and veml6030 subprojects in the mix.exs file. We'll do the preponderance of the configuration and dependency management in the firmware project and as much of the rest of the project code as possible in the dependency projects.

With a good sense as to how the Nerves application will be structured, we'll need to assemble our weather station before we can have it executing any code. Let's look into the hardware required for this project and how it all looks once it's assembled.

Assembling the Weather Station

To build our IoT sensor hub, we'll need some sensors and a way to attach those sensors to the board. Typically, sensors are tiny chips that must be attached to circuit boards called breakout boards. These boards are beyond the scope of this book, but the good news is that there are plenty of interfaces and pre-built breakout boards at our disposal. We're going to use a solderless connect interface called Qwiic Connect System. Using the Qwiic Connect System, we'll be able to attach I2C[12]-compatible sensors to our Nerves IoT and get up and running in record time.

What Is I2C?

 Inter-Integrated Circuit (or I2C for short) is a communication protocol that allows us to connect multiple external devices to one or more host devices. The external devices (in our case sensors) can all be daisy-chained together and communicate with the host device over the same data bus.

You can set up your Raspberry Pi to support Qwiic Connect sensors by either buying an easy to use header HAT,[13] or if you are feeling more DIY, you can also solder your own Qwiic Connect SHIM[14] onto your Raspberry Pi. You can also buy some sensors on breakout boards that already have Qwiic Connect headers available, so that you can assemble the project quickly. Let's take a look at what's needed to build the weather station.

---

11. https://hexdocs.pm/mix/Mix.Tasks.Deps.html
12. https://learn.sparkfun.com/tutorials/i2c/all
13. https://www.sparkfun.com/products/15945
14. https://www.sparkfun.com/products/15794

Gathering the Hardware

A maker store called SparkFun is a good place to get sensors, so we'll build our whole project list from that site. These products come and go, and prices frequently change, so be sure to shop around for the best deal. We're going to need a Raspberry Pi along with some supporting hardware, an environment sensor, an air quality sensor, an ambient light sensor and a way to connect it all using the Qwiic interface:

Raspberry Pi Zero W with headers[15]

> The computer that will serve as our IoT sensor hub. You're not limited to only the Raspberry Pi Zero W, as Nerves supports a wide array of embedded devices, but this one is the most cost-effective device out there. If you get the Raspberry Pi with the headers pre-soldered then you can easily connect your sensors to the 2x20 rows or pins.

Qwiic pHAT v2.0 for Raspberry Pi[16]

> This board will allow you to easily connect Qwiic Connect breakout boards to your Raspberry Pi and communicate with other I2C devices. As previously mentioned, you can opt for the SparkFun Qwiic SHIM for Raspberry Pi instead if you're looking for something cheaper and more DIY.

VEML6030 Light Sensor (Qwiic)[17]

> A device that can detect light and connect to the Raspberry Pi over a standardized interface called I2C.

BME680 Environmental Sensor[18]

> A sensor to measure temperature, humidity, and barometric pressure, and connect to the Raspberry Pi over I2C.

SGP30 Air Quality Sensor[19]

> A sensor to detect air quality that can be chained to other sensors and connected to the Raspberry Pi over I2C.

Qwiic Connection Cables[20]

> If your sensors don't come with these cables, get a few. They are usually under a dollar.

---

15. https://www.sparkfun.com/products/15470
16. https://www.sparkfun.com/products/15945
17. https://www.sparkfun.com/products/15436
18. https://www.sparkfun.com/products/16466
19. https://www.sparkfun.com/products/16531
20. https://www.sparkfun.com/products/14426

MicroUSB connection cables[21]

You'll need to make a wired connection from your Raspberry Pi to your computer to configure the network and to power the device.

4GB+ MicroSD card[22]

You'll need a MicroSD card to store the Nerves firmware for your Raspberry Pi. Anything with 4GB of capacity or greater will do fine.

Other I2C sensors[23]

You might want to connect other sensors over I2C as well. Use a Qwiic connector, if possible, to connect to other sensors on the chain.

Assembling the Sensor Hub

The following picture shows all the components that we'll be working with. From breakout boards to Qwiic cables to the Raspberry Pi itself along with the Qwiic Connect HAT, this is everything that we'll need to capture environmental data.

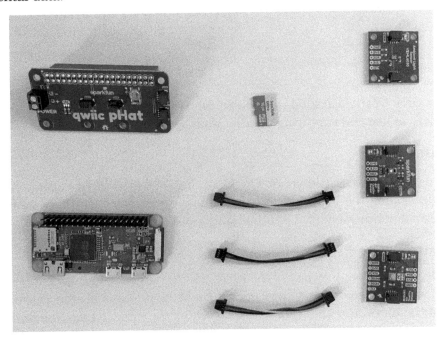

---

21. https://www.sparkfun.com/products/10215
22. https://www.sparkfun.com/products/15051
23. https://www.sparkfun.com/qwiic

Putting together the sensor hub should feel like putting together a collection of Lego pieces. The Qwiic Connect HAT should slide nicely on top of the Raspberry Pi header pins, and from there you can daisy chain all the breakout boards together (in no particular order, as I2C doesn't dictate any particular orientation), as shown in the following picture.

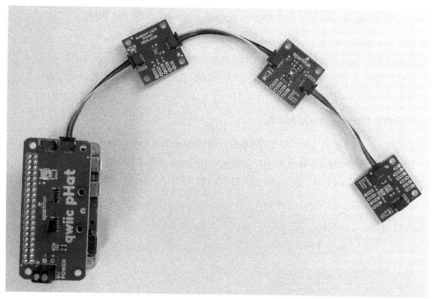

With your weather station assembled, you're ready to start putting those sensors to use and capturing weather data.

Your Turn

In this chapter we talked about why Elixir and Nerves are an excellent choice for developing the IoT application and how we plan to use those tools to develop a network-enabled weather station.

What You Built

While we didn't get into any of the Elixir and Nerves code in this chapter, we discussed what components we'll need to build the project and even assembled all of the hardware. This sets us up nicely for the next chapter, where we'll be writing all of our Nerves-specific code and will even pull sensor data directly from our breakout boards.

Why It Matters

The material covered in this chapter was important in that it laid the foundation for what we'll be building, how these concepts align with the real world, and how we might approach similar problems in the future given our toolbox of Elixir and Nerves.

What's Next

In the next chapter, we'll be diving head first into the code that will power our device. We'll take all of the architecture-related concepts that we covered in this chapter and will be putting them into practice in our Nerves sensor hub poncho project.

CHAPTER 2

Wirelessly Reading Sensor Data

Now that we have a clear idea what we'll be building and how we'll be approaching the problem, let's start off by creating a brand-new Nerves project and getting on to our home wireless network.

Creating a Network-Enabled Sensor Hub Project

To get things rolling we're going to start by generating a new Nerves project, then burning the firmware to the device, and finally connecting to the device via SSH. Once we connect to the device over SSH, we'll be able to configure the network settings.

Installing the Nerves Project Generator

 Prior to using the Nerves project generator and the Nerves CLI tools, you'll need to make sure that your workstation is correctly set up. Be sure to visit the installation instructions on HexDocs to get everything installed for your particular platform (MacOS, Linux, or Windows).[1]

Once you have the Nerves tooling in place, we'll start out by creating a top-level directory that will house our poncho project. After that, we'll change directories into that top-level poncho directory and create a new Nerves project, as we discussed in the previous chapter:

```
$ mkdir sensor_hub_poncho
$ cd sensor_hub_poncho
```

---

1. https://hexdocs.pm/nerves/installation.html

```
$ mix nerves.new sensor_hub
* creating sensor_hub/config/config.exs
* creating sensor_hub/config/host.exs
* creating sensor_hub/config/target.exs
* creating ...

Your Nerves project was created successfully.

...
```

For the next series of commands, you need to tell Nerves what "target" you're acting upon. In other words, Nerves needs to know what platform it's building a firmware for. In this book we'll be using a Raspberry Pi 0 Wireless, but if you happen to have another Raspberry Pi model or even a BeagleBone, be sure to check out the supported target list.[2]

Since you'll be performing a few commands in rapid succession, it will be easier if you export the target environment variable for the duration of your terminal session. In addition, you'll also need to have your MicroSD card inserted into your workstation so that you can burn your first Nerves firmware to the MicroSD card (make sure that the correct MicroSD card is being used during the mix burn step if there are multiple MicroSD cards attached to your workstation).

```
$ cd sensor_hub
$ export MIX_TARGET=rpi0

$ mix deps.get
...

$ mix firmware
...

$ mix burn
==> nerves
==> sensor_hub

Nerves environment
  MIX_TARGET:   rpi0
  MIX_ENV:      dev

Use 14.84 GiB memory card found at /dev/rdisk2? [Yn] y
100% [====================================] 33.32 MB in / 35.76 MB out
Success!
Elapsed time: 4.765 s
```

---

2.   https://hexdocs.pm/nerves/targets.html#supported-targets-and-systems

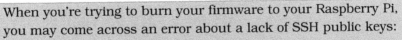

Setting up Device SSH Keys

When you're trying to burn your firmware to your Raspberry Pi, you may come across an error about a lack of SSH public keys:

```
** (Mix) No SSH public keys found in ~/.ssh. An ssh authorized ...
log into the Nerves device and update firmware on it using ssh.
See your project's config.exs for this error message.
```

Be sure to follow the directions laid out in the Nerves documentation to set up your SSH keys.[3]

And just like that, you have a super-slim Erlang-based firmware burned onto your MicroSD card with your vanilla Elixir Nerves application! You can now insert your MicroSD card into the Raspberry Pi and SSH over to it once it has started up (you can leave your sensor hub assembled as it was at the end of Chapter 1). Be sure that you're also using a USB data cable and not a power-only cable:

```
$ ssh nerves.local
...
iex(1)> hostname()
"nerves-dc74"

iex(2)> exit()
Connection to nerves.local closed.
```

Once you know the hostname of your Nerves device, you can also SSH into it by using its name. For example, you could run ssh nerves-dc74.local instead of ssh nerves.local and also connect to the Nerves device (be sure to replace nerves-dc74.local with whatever the hostname of your device is).

As a side note, be sure to connect the USB cable from your workstation to the port marked USB and not PWR IN. The reason for this is that the port marked with USB enables you to have a virtual Ethernet network connection with the Raspberry Pi, which is how you are able to SSH into the Raspberry Pi. The first image on page 18 highlights which MicroUSB port you should use to connect to your device to enable the virtual Ethernet connection:

You should now be up up and running if you're able to SSH into the device! If you'd like to terminate your IEx session, you can run the exit() function as shown in the second image on page 18.

Now let's get to work on understanding networking on the Nerves-powered device and how to configure our wireless networking interface.

---

3. https://hexdocs.pm/nerves_firmware_ssh/readme.html#device-keys

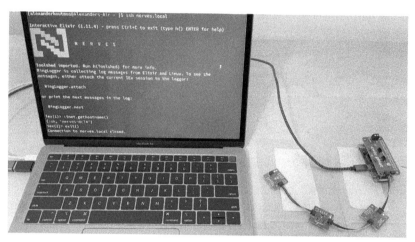

Getting on to the Network

As previously mentioned, you're already networking with your project but through a wired connection over the USB cable. Shelling into the Raspberry Pi and updating firmware both use this USB network. Let's take a look at the configuration that makes USB networking possible. Open up sensor_hub/config/target.exs:

```
config :vintage_net,
  regulatory_domain: "US",
  config: [
    {"usb0", %{type: VintageNetDirect}},
    {"eth0",
     %{
       type: VintageNetEthernet,
       ipv4: %{method: :dhcp}
     }},
    {"wlan0", %{type: VintageNetWiFi}}
  ]
```

This config entry establishes the configuration for all of the networking needs related to your project. The line of interest in this case is {"usb0", %{type: Vintage NetDirect}}. The name usb0 is important in that it is the network interface name that the Linux kernel uses to access the virtual USB Ethernet. The second value in that first tuple is a configuration map that VintageNet[4] uses under the hood to configure that particular network interface. The VintageNetDirect module configures a direct connection over usb0. While it accepts additional options, the defaults work well for point-to-point Ethernet links like the one we're using.

Configuring the Wireless Network

As our project evolves, we'll want to migrate away from using the USB cable for networking and instead leverage WiFi. This will let us plug the Raspberry Pi Zero W into a USB wall power supply or battery pack and run disconnected from our workstation. You can either create a new access point on the Raspberry Pi (that is, the Raspberry Pi is hosting its own WiFi network) or connect to an existing access point.

We're going to take the second option and connect to an existing access point. If you want, you can configure the network connection directly in the target.exs configuration file. While that approach will work, you need to be careful when it's time to check that configuration file into source control. Instead, we'll take another approach—we'll shell into the Raspberry Pi and build a persistent connection.

Before setting up our wireless network interface, we should do one more thing. If we want our device to access the internet, we'll need to add :inets to our extra_applications list in mix.exs. Then, with our MIX_TARGET=rpi0 environment variable exported, call mix firmware and mix upload. This new mix upload command will perform an upgrade of the firmware on the Nerves device, all over SSH. Now, it's time to configure the wireless network to make full use of our Raspberry Pi's wireless adapter.

Build a Persistent Wireless Configuration

To connect to a wireless access point, you'll need to specify a configuration, either in target.exs or by using VintageNet via remote IEx. To keep the configuration out of source control, let's add it via shelling into the Raspberry Pi. With your Raspberry Pi hooked up via the USB cable, shell into your system:

---

4. https://github.com/nerves-networking/vintage_net

```
ssh nerves.local
...

iex(1)> VintageNet.configure("wlan0", %{
...(1)>    type: VintageNetWiFi,
...(1)>    vintage_net_wifi: %{
...(1)>      networks: [%{
...(1)>        key_mgmt: :wpa_psk,
...(1)>        ssid: "<YOUR NETWORK NAME>",
...(1)>        psk: "<YOUR WIRELESS PASSWORD>"
...(1)>      }]
...(1)>    },
...(1)>    ipv4: %{method: :dhcp}
...(1)>  })
:ok
```

After we shell into the device and configure the network, VintageNet will auto-matically save that configuration to persistent storage on the Raspberry Pi. Alternatively, you can use the abbreviated API that takes this form:

```
iex> VintageNetWiFi.quick_configure("<NETWORK NAME>", "<WIRELESS PASSWORD>")
```

With either approach, we can make use of IEx to make sure that the file is on the device:

```
iex> ls "/data/vintage_net/wlan0"
/data/vintage_net/wlan0
```

We're close to having our device decoupled from our workstation. All we need to do now is disconnect the USB cable from the USB port, and then hook it up to the *other* MicroUSB on the Raspberry Pi (the one marked PWR IN). The following image highlights the MicroUSB port that you should use for only powering the device:

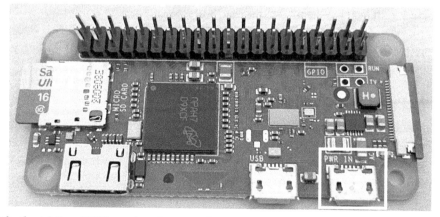

With that MicroUSB port only providing power to the device, we'll be able to check and see whether the wireless network is configured properly. You can either plug the other end of the USB cable back into your computer to provide power or use a USB power adapter. After giving the Raspberry Pi a minute or so to boot up, you should be able to remote shell back into the Raspberry Pi by running ssh nerves.local in the terminal. Once connected to the Nerves device, run the following to ensure you have internet connectivity:

```
iex(1)> ping "akoutmos.com"
Press enter to stop
Response from akoutmos.com (52.73.87.228): time=31.992ms
Response from akoutmos.com (52.73.87.228): time=35.048ms
ctrl-c
```

Brilliant! We're connected to our wireless network and to the Internet. Now, enter this command to get info about the connection:

```
iex(1)> VintageNet.info()
VintageNet 0.9.3

All interfaces:        ["lo", "usb0", "wlan0"]
Available interfaces: ["wlan0"]

Interface eth0
  ...

Interface usb0
  ...

Interface wlan0
  Type: VintageNetWiFi
  Present: true
  State: :configured (28.7 s)
  Connection: :internet (21.2 s)
  Addresses: 192.168.1.52/24, fe80::ba27:ebff:fecb:8921/64
  Configuration:
    %{
      ipv4: %{method: :dhcp},
      type: VintageNetWiFi,
      vintage_net_wifi: %{
        networks: [
          %{
            key_mgmt: :wpa_psk,
            mode: :infrastructure,
            psk: "....",
            ssid: "<YOUR NETWORK NAME>"
          }
        ]
      }
    }
```

As you can see with the wlan0 entry, it's present, it has an IP address, and it's connected. Since you know the IP address is 192.168.1.52, you can even ping it from your host. If you ever have any trouble, be sure to check out the tools available to you via h Toolshed so that you can debug your network connection troubles. With our device now hooked up to the wireless network, it's time to start reading data from our array of sensors.

Capturing Sensor Data

The interface we'll be working with is Inter Integrated Circuit Protocol, or I2C for short. It's a protocol used to connect two integrated circuits. The standard is a communications bus, meaning you can use it to attach multiple devices connected together. You can have one controller connected to multiple peripherals, as we will, or even multiple controllers connected to multiple peripherals, which comes in handy when you're building more complex circuits.

Broadly, the protocol uses a system clock and voltages to send values between devices. Each I2C device has an address, and we'll typically send a request and receive a response. We'll read the spec sheet to find out exactly which values you'll need to interact with the light sensor.

Let's download the spec sheet and see what it says.

Reading the Spec Sheet

If you look closely at our light sensor board, you'll see that the underlying sensor is the VEML6030 ambient light sensor. The chip maker has two important PDFs, a spec sheet[5] and an application guide,[6] that define how to interact with the sensor. All the details we need are there, but there's a ton of information, and if you don't work with hardware every day, some of it might be confusing. Sparkfun also has a guide[7] with much of the information we need to hook it up.

We're going to express values in hexadecimal because it makes dealing with the various bytes and bits the sensor needs easier. In Elixir, a hex value is written in the form 0xff. The sensor has an address of 0x48. It also has a feature called an interrupt that you can configure, but we won't use it. If we wanted to, we could trip the interrupt to fire when the sensor detects levels above or below a certain threshold.

Instead, we're interested in reading the value from the sensor and reporting it over to our Phoenix API server. The sensor supports various sensitivities,

---

5.   https://www.vishay.com/docs/84366/veml6030.pdf
6.   https://www.vishay.com/docs/84367/designingveml6030.pdf
7.   https://learn.sparkfun.com/tutorials/qwiic-ambient-light-sensor-veml6030-hookup-guide

called gains. The possible values are 1, 2, 1/8, and 1/4 , and we're going to work with a fairly low setting of 1/4 based on the advice in the Sparkfun guide.

The sensor also needs an integration time setting. The trade-off is that a higher time takes longer but lets you measure dimmer light levels. The possible values are 800, 400, 200, 100, 50, and 25. We're going to use 100. That means we'll need to wait at least a tenth of a second between readings.

Remember, the readings we get back will have several different resolutions, and that means the sensor value we get back will have to take that resolution into account. There's a conversion table in both the spec sheet[8] and the Sparkfun guide.[9]

Keep your spec sheets handy, because it's time to use them to interact with the device.

Fetching Sensor Data via Circuits.I2C

Now, it's time to power up the Raspberry Pi and update our code a bit. You've probably noticed any time your device has power, all the connected sensors have a bright red LED that lights up. If you read the label right above the LED, you'll see it's marked as PWR. This let's us know the sensor is operational and is ready to interface with our Raspberry Pi. If everything is correctly assembled, your weather station should look like so when it has power:

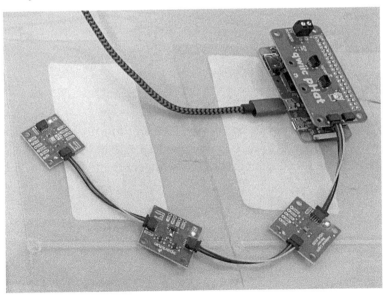

---

8.   https://www.vishay.com/docs/84366/veml6030.pdf

9.   https://learn.sparkfun.com/tutorials/qwiic-ambient-light-sensor-veml6030-hookup-guide

Let's open up mix.exs and add an additional dependency to our project. This dependency allows us to interface with I2C sensors and is maintained by the Elixir Circuits GitHub organization.[10] All we need to do is add the following to our sensor_hub/mix.exs file:

```
defp deps do
  [
    # Dependencies for all targets
    ...
    {:circuits_i2c, "~> 0.3.8"},

    # Dependencies for all targets except :host
    ...

    # Dependencies for specific targets
    ...
  ]
end
```

With that in place, run the following commands in your terminal so that you can fetch your new dependency, create a new firmware, upload the new firmware to your Raspberry Pi, and lastly, connect to the device after it has rebooted (make sure you're in the sensor_hub poncho project and MIX_TARGET=rpi0 has been exported):

```
$ mix deps.get
...

$ mix firmware
...

$ mix upload
...

$ ssh nerves.local
...
```

Now that we're connected to the device (SSHing into the device may take a minute while the new firmware is swapped for the old firmware), let's alias the Circuits.I2C library and see what devices are connected:

```
iex(1)> alias Circuits.I2C
Circuits.I2C

iex(2)> I2C.detect_devices()
Devices on I2C bus "i2c-1":
 * 72   (0x48)
 * 88   (0x58)
 * 119  (0x77)

3 devices detected on 1 I2C buses
```

---

10. https://github.com/elixir-circuits/circuits_i2c

Excellent. You'll notice that there are three devices connected to the I2C bus (one entry for each of our sensors). Let's play with the VEML6030 ambient light sensor (which is device 0x48) from the IEx REPL to see if we can extract a measurement reading.

A common I2C pattern is to write to a device and register and then read the value that comes back. The Elixir Circuits library wraps up this write/read pattern in a single function call. Reading the default setting will show how the pattern works. The write_read!/4 function takes an i2c reference, a device address, a register, and the number of bytes to read. The result will come back as a 16-bit value, with the least significant byte first (also known as little-endian). Since typical Elixir returns the most significant byte first (big-endian), we'll need to tweak the result. Let's use the command to fetch register 0 to see what the value is:

```
iex(1)> alias Circuits.I2C
Circuits.I2C

iex(2)> sensor = 0x48
72

iex(3)> command = <<0>>
<<0>>

iex(4)> byte_size = 2
2

iex(5)> {:ok, i2c_ref} = I2C.open("i2c-1")
{:ok, #Reference<0.1635386997.268828675.62058>}

iex(6)> <<value::little-16>> = I2C.write_read!(i2c_ref, sensor, command, 2)
<<1, 0>>

iex(7)> value |> inspect(base: :binary)
"0b1"
```

We start off by setting up a few constants for use throughout our REPL session. Then we fetch a value from our I2C-enabled light sensor, and we finally convert the value to binary to make it easier to compare to the spec sheet. As you can see, the value of 0b1 means that the device is turned off.

Let's fix that.

The spec sheet tells us that we need to set bits 11 and 12 to 0b11 to set the gain. The integration time setting is in bits 6 through 9, and we need to set those all to 0. That's convenient. We also need to set bit 0 to 0, to power things up. We can set that value using a 16-bit binary number with the 0bxxxx syntax, and then write to register 0 to open the sensor, like this:

```
iex(8)> config = 0b0001100000000000
6144

iex(9)> Circuits.I2C.write(i2c_ref, sensor, <<0, config::little-16>>)
:ok

iex(10)> <<value::little-16>> = I2C.write_read!(i2c_ref, sensor, command, 2)
<<0, 24>>

iex(11)> value |> inspect(base: :binary)
"0b1100000000000"
```

We specify the sensor by building a 16-bit number using the bits on the spec sheet. We get back an :ok, and then we read the register and get the same number back. It's open, and it's configured just as we expect it to be.

With that done, our sensor is ready to use. The spec sheet says the light reading is in register 4, so let's read that value:

```
iex(12)> light_reading = 4
4

iex(13)> <<value::little-16>> =
...(13)>   Circuits.I2C.write_read!(i2c_ref, sensor, <<light_reading>>, 2)

iex(14)> value
440
```

We get a value back. Now, put your hand over the sensor so that there's less light and see if the value is lower:

```
iex(15)> <<value::little-16>> =
...(15)>   Circuits.I2C.write_read!(i2c_ref, sensor, <<light_reading>>, 2)

iex(16)> value
40
```

It's lower, so the sensor is working as expected! But what do the numbers mean? It turns out that we need to apply a conversion factor based on our settings. If you read the application guide under the heading "Translating ALS Measurement Results into a Decimal Value," you'll find a factor to use for converting to lumens, a measurement of light for a rough conversion. In some circumstances, such as working through tinted glass or very bright lights, you'll need to apply a correction formula, but the rough values will fit our purposes just fine. Our conversion factor is 0.2304. Let's write a quick function in our IEx session to read, extract the value, and convert our measurements to lumens:

```
iex(17)> measure_light = fn i2c, address ->
...(17)>   <<value::little-16>> = I2C.write_read!(i2c, address, <<4>>, 2)
...(17)>   value * 0.2304
...(17)> end
```

```
#Function<43.97283095/2 in :erl_eval.expr/5>
iex(16)> measure_light.(i2c_ref, sensor)
57.6
```

This function takes the data we've accumulated so far and rolls it up into a convenient function. Now, drag your family or co-workers around the house, trying your sensor in a variety of locations:

```
iex(18)> IO.puts("In the shade: #{measure_light.(i2c_ref, sensor)}")
In the shade: 64.512
:ok

iex(19)> IO.puts("In the sunlight: #{measure_light.(i2c_ref, sensor)}")
In the sunlight: 5339.9808
:ok

iex(20)> IO.puts("In the closet: #{measure_light.(i2c_ref, sensor)}")
In the closet: 1.3824
:ok
```

Marvelous! The sensor is reporting pretty broad and sensitive responses, depending on the surrounding environment. Now we have a working sensor for the hub. That's a good start, and you also have a second sensor to try out. It's time for you to apply these techniques.

Your Turn

In this chapter, you found out why Elixir and Nerves provide an excellent platform for creating network-enabled IoT devices. By leveraging the provided Nerves networking tools, you were able to connect to your device, remotely push firmware updates, and read live sensor data. All of this was possible thanks to the solid foundation provided to us by Elixir and Nerves. This allowed us to focus on the core of our problem as opposed to worrying about the lower-level details.

What You Built

You created the first poncho project in the application using the Nerves mix nerves.new command. With your newly created project, you were able to burn your firmware to the device and SSH into it using the default wired connection over a USB cable. With an active IEx REPL session, you were able to run a few commands and establish a wireless connection. Thanks to the Nerves VintageNet tooling, your network configuration was automatically saved so that whenever the device reboots, it can connect to the wireless network without any manual intervention from you.

Sensors in Nerves typically use the I2C interface, and have broadly different interfaces to access them. The typical model for working with them is to write to one register to issue a command and read from another register as a response. The Circuits.I2C library has some functions to make that process easier, as you saw with the light sensor.

Why It Matters

This was your first pass at getting things up and running with Nerves and our Raspberry Pi. You were able to accomplish a lot toward your end goal of an environmental sensor hub with only a little configuration and some initial set up. The important takeaway here is that you now have a development workflow established where you can make code changes on your workstation and then, over the wireless network, flash your device with an updated firmware. This is one of the main benefits of Elixir and Nerves, in that you can rapidly build and test your devices without out much ceremony.

What's Next

This chapter got you ready for the work you'll do in the next chapter, where you'll build a core and boundary layer in order to set up your sensor hub to publish metrics to your Phoenix API server. You'll also set up the remainder of your sensors so that you can get a holistic view of the environment using your Nerves-powered Raspberry Pi.

CHAPTER 3

Aggregating Sensor Data

Now that we have a productive development workflow in place, it's time to get our project into a production-level state. While connecting to our Nerves device over the wireless network is a great way to validate that things are working as expected, our goal is to get things working autonomously so that sensor data can be aggregated and published automatically when the device powers on. To achieve that, we'll need to wrap each sensor in its own GenServer[1] and add it to our application supervision tree. That way we can leverage the fault-tolerant and reliable properties of Erlang and OTP for our IoT weather station.

Wrapping Sensors in GenServers

From the previous chapter, you already know the VEML6030 sensor works well with the Circuits.I2C interface, and you can to interact with it by writing I2C commands and reading the response. While users could call the sensor straight from I2C, it would be nice to shield them from many tedious I2C details. The spec sheets have details, such as configuration settings and conversion factors, that our users shouldn't have to know.

While we could put the GenServer wrapper directly in the sensor_hub project, it would be better for our architecture to instead lean on the poncho project structure and split this out into its own separate project. We can then lean on Mix's ability to resolve path dependencies and use our veml6030 project directly from our sensor_hub firmware project.

With that said, let's build a skinny GenServer wrapper. Later, if there's a need and if we build up a critical mass of features, we can extract it into a published

---

1.   https://hexdocs.pm/elixir/GenServer.html

Hex package so the community can share the costs and benefits of working with the project.

Navigate to the sensor_hub_poncho directory and create a new project:

```
$ mix new veml6030
$ cd veml6030
```

After the project is created, update the mix.exs file in the veml6030 subproject to have the following dependencies (given that we need to communicate with an I2C sensor, we'll need the Circuits.I2C library):

```
defp deps do
  [
    {:circuits_i2c, "~> 0.3.8"}
  ]
end
```

The veml6030 subproject will be a basic dependency project. Supervision of the VEML6030 GenServer will happen later in the sensor_hub firmware project, which is responsible for starting and configuring those dependencies.

Since this project is complex, we'll start our code where development is simple and predictable. We'll implement the hardest part of the project within the core, the layer that builds commands for the sensor.

Build a Core from the Spec Sheet

Much of a typical hardware project is dedicated to moving bits and bytes around in various configurations. Elixir bitstrings make quick work of such tasks. The most difficult part of working with our light sensor is to send a command that represents the configuration of the data we want to read. Pretty much everything else is sending and receiving trivial one-byte commands via I2C and reading the result.

 Bruce says:
Construct |> Reduce |> Convert

CRC is a way to think about composing functions that mirror an Elixir pipe. Constructors at the head of the pipe transform convenient inputs to a central type convenient for consumption. Reducers in the middle of a pipe do a tiny bit of work. Converters at the end of a pipe transform the central data type into data that's convenient for consumption. See this blog post for more information.[a]

a. https://redrapids.medium.com/learning-elixir-its-all-reduce-204d05f52ee7

Let's create a file located at lib/veml6030/config.ex that represents an abstract configuration. In it, you'll build a constructor to build the individual components, and a converter to convert to the configuration command you'll eventually send. After you create a configuration, you'll never change it, so there's no need for a reducer.

Let's start of by creating the lib/veml6030/config.ex file one chunk at a time. Add the following contents to lib/veml6030/config.ex:

```
defmodule VEML6030.Config do
  defstruct [
    gain: :gain_1_4th,
    int_time: :it_100_ms,
    shutdown: false,
    interrupt: false
  ]

  def new, do: struct(__MODULE__)
  def new(opts), do: struct(__MODULE__, opts)
end
```

This new VEML6030.Config module will contain the configuration settings, including settings for the integration time via :int_time and gain via :gain. We also include shutdown and interrupt settings in our struct since these features can be supported through I2C.

With the constructor out of the way, let's move on to the converter. It will use bit strings and small custom functions that plug in individual values, straight from the spec sheet, like this:

```
defmodule VEML6030.Config do
  ...

  def to_integer(config) do
    reserved = 0
    persistence_protect = 0

    <<integer::16>> = <<
      reserved::3,
      gain(config.gain)::2,
      reserved::1,
      int_time(config.int_time)::4,
      persistence_protect::2,
      reserved::2,
      interrupt(config.interrupt)::1,
      shutdown(config.shutdown)::1
    >>

    integer
  end
end
```

That's the overall format of the command. You can see that it translates directly to the spec sheet, and that makes us happy. Now we'll build short functions using pattern matching to handle each of the settings, like this:

```elixir
defmodule VEML6030.Config do
  ...

  defp gain(:gain_1x), do: 0b0
  defp gain(:gain_2x), do: 0b01
  defp gain(:gain_1_8th), do: 0b10
  defp gain(:gain_1_4th), do: 0b11
  defp gain(:gain_default), do: 0b11

  defp int_time(:it_25_ms), do: 0b1100
  defp int_time(:it_50_ms), do: 0b1000
  defp int_time(:it_100_ms), do: 0b0000
  defp int_time(:it_200_ms), do: 0b0001
  defp int_time(:it_400_ms), do: 0b0010
  defp int_time(:it_800_ms), do: 0b0011
  defp int_time(:it_default), do: 0b0000

  defp shutdown(true), do: 1
  defp shutdown(_), do: 0

  defp interrupt(true), do: 1
  defp interrupt(_), do: 0
end
```

These tiny functions are super-simple. Each one converts a configuration setting to the bits from the spec sheet that trigger the command. All that remains is a conversion factor that the user will apply to each measurement. Since the conversion factor is based on the configuration, we'll put it in a module attribute and translate it within the configuration, like this:

```elixir
defmodule VEML6030.Config do
  # There's more to this lumens factor map. For the full listing see
  # the nerves_code/veml6030/lib/veml6030/config.ex file in the
  # https://github.com/akoutmos/nerves_weather_station repo.
  @to_lumens_factor %{
    {:it_800_ms, :gain_2x} => 0.0036,
    {:it_800_ms, :gain_1x} => 0.0072,
    {:it_800_ms, :gain_1_4th} => 0.0288,
    {:it_800_ms, :gain_1_8th} => 0.0576,
    ...
  }

  def to_lumens(config, measurement) do
    @to_lumens_factor[{config.int_time, config.gain}] * measurement
  end
end
```

This table is pretty long, so we've omitted much of it. If you would like to see the whole table, check out the GitHub project in the comment. Let's take it for a quick test drive on our host development machine by running iex -S mix:

```
iex(1) ▶ VEML6030.Config.new()
%VEML6030.Config{
  gain: :gain_1_4th,
  int_time: :it_100_ms,
  interrupt: false,
  shutdown: false
}
iex(2) ▶ VEML6030.Config.new() |>
...(2) ▶ VEML6030.Config.to_integer() |>
...(2) ▶ inspect(base: :hex)
"0x1800"

iex(3) ▶ [gain: :gain_1x] |>
...(3) ▶ VEML6030.Config.new() |>
...(3) ▶ VEML6030.Config.to_integer() |>
...(3) ▶ inspect(base: :hex)
"0x0"
```

If you check the VEML6030 spec sheet, you'll see that these values are all correct. Now we need to consume these functions in a hardware layer.

Create Boundary Hardware Layer

The lib/veml6030/config.ex file does a good job of representing a configuration, both abstractly and in bytes. It also shields the user from tedious ceremony by choosing sane defaults, although it's only a solution to part of the problem of interacting with this sensor, as it doesn't yet interact with the hardware.

We already know how to open the device using Circuits.I2C, but there's also another function called discover_one!/1 that we can use to simplify the experience. Create a file called lib/veml6030/comm.ex and start by adding this:

```
defmodule VEML6030.Comm do
  alias Circuits.I2C
  alias VEML6030.Config

  @light_register <<4>>
end
```

We alias the core we made and the Circuits.I2C hardware interface. Now it's time to make use of that new discover_one!/1 function:

```
defmodule VEML6030.Comm do
  ...
```

```
    def discover(possible_addresses \\ [0x10, 0x48]) do
      I2C.discover_one!(possible_addresses)
    end
end
```

Different IoT devices and different sensors present different combinations of buses and addresses. By narrowing the list of addresses to only a few possibilities, the I2C library can scan for a working bus and sensor. Then we can present an API to open the sensor, raising on a failure, like this:

```
defmodule VEML6030.Comm do
  ...

  def open(bus_name) do
    {:ok, i2c} = I2C.open(bus_name)

    i2c
  end
end
```

It's just an I2C.open/1 and a pattern match. Since we have an interface that can build the configuration command, writing the configuration becomes a trivial I2C write:

```
defmodule VEML6030.Comm do
  ...

  def write_config(configuration, i2c, sensor) do
    command = Config.to_integer(configuration)

    I2C.write(i2c, sensor, <<0, command::little-16>>)
  end
end
```

We take the configuration, convert it to a command, and write it. Notice that the command register is 0, and the spec sheet says commands must be sent as a 16-bit little-endian encoding. In Elixir, we do so by expressing 16-bit bytes—with the least significant byte first—using the ::little-16 encoding.

Now we can read and unpack the result by writing the register name and converting the result to lumens.

```
defmodule VEML6030.Comm do
  ...

  def read(i2c, sensor, configuration) do
    <<value::little-16>> =
      I2C.write_read!(i2c, sensor, @light_register, 2)

    Config.to_lumens(configuration, value)
  end
end
```

That all works fine. We'll wait to exercise this code until we're able to stand up the GenServer and reference it from the sensor_hub project. For now, it's time to move on.

Wrap the Core and Hardware in a GenServer

It's customary to wrap sensors in OTP GenServers because there are often requirements to read from them periodically to get consistent readings. In our case, the wrapping is also practical because the code must wait between readings. The amount of time is specified in the sensor's spec sheet. By periodically taking measurements every second, we'll be fine. Let's build a GenServer to do so.

Since the GenServer will also double as the API for the whole project, we can put the GenServer in the top-level lib/veml6030.ex file. Go ahead and delete the module code created by mix new and start with the GenServer boilerplate:

```elixir
defmodule VEML6030 do
  use GenServer

  require Logger

  alias VEML6030.{Comm, Config}
end
```

Here we use, require, and alias various dependencies as needed. Next, add two versions of the init/1 GenServer callback, like so:

```elixir
defmodule VEML6030 do
  ...

  @impl true
  def init(%{address: address, i2c_bus_name: bus_name} = args) do
    i2c = Comm.open(bus_name)

    config =
      args
      |> Map.take([:gain, :int_time, :shutdown, :interrupt])
      |> Config.new()

    Comm.write_config(config, i2c, address)
    :timer.send_interval(1_000, :measure)

    state = %{
      i2c: i2c,
      address: address,
      config: config,
      last_reading: :no_reading
    }

    {:ok, state}
  end
end
```

This version creates a configuration and opens the I2C bus. It also sends a periodic message to itself at a one-second interval. Then it sets the state of the GenServer.

Now we can write an alternative init to discover the address and bus if they're not set, like this:

```elixir
defmodule VEML6030 do
  ...

  def init(args) do
    {bus_name, address} = Comm.discover()
    transport = "bus: #{bus_name}, address: #{address}"

    Logger.info("Starting VEML6030. Please specify an address and a bus.")
    Logger.info("Starting on " <> transport)

    defaults =
      args
      |> Map.put(:address, address)
      |> Map.put(:i2c_bus_name, bus_name)

    init(defaults)
  end
end
```

If either the address or bus is not there, the code automatically discovers the sensor, creates a set of default configurations, and calls our prior init/1 implementation. Next, we'll write the GenServer callbacks to handle our :measure message that's sent at a regular one-second interval and the :get_measurement message that returns the GenServer's last measurement reading:

```elixir
defmodule VEML6030 do
  ...

  @impl true
  def handle_info(
        :measure,
        %{i2c: i2c, address: address, config: config} = state
      ) do
    last_reading = Comm.read(i2c, address, config)
    updated_with_reading = %{state | last_reading: last_reading}

    {:noreply, updated_with_reading}
  end

  @impl true
  def handle_call(:get_measurement, _from, state) do
    {:reply, state.last_reading, state}
  end
end
```

The handle_info/2 callback, which handles the :measure message, updates the :last_reading key in our GenServer state based on the response from our Comm module. This updated GenServer state now contains the latest light measurement from the VEML6030 sensor.

To make this data easily accessible, we'll also want to add a couple of public API functions:

```
defmodule VEML6030 do
  ...

  def start_link(options \\ %{}) do
    GenServer.start_link(__MODULE__, options, name: __MODULE__)
  end

  def get_measurement do
    GenServer.call(__MODULE__, :get_measurement)
  end

  ...
end
```

Our start_link/1 function wraps the GenServer start_link/3 function and sets the name of the GenServer process to VEML6030 (the name of the module). The get_measurement/0 function is another GenServer wrapper that makes it easier to fetch the light measurement from the GenServer process. Since we gave our GenServer the :name option, we can call it directly via __MODULE__ (think of this as a singleton process).

Let's go ahead and test all of this out on our Raspberry Pi. Open up the mix.exs file in the sensor_hub subproject and add the following to your dependencies:

```
defp deps do
  [
    ...

    # Dependencies for all targets except :host
    {:veml6030, path: "../veml6030", targets: @all_targets},
    ...
  ]
end
```

Then, in a terminal session, run the following commands to burn the firmware to the device (assuming that your Raspberry Pi is connected to the wireless network and that you have exported MIX_TARGET=rpi0):

```
$ cd sensor_hub
$ mix deps.get
...

$ mix firmware
...
```

```
$ mix upload
...

# You may have to wait a little until the device reboots
$ ssh nerves.local
...
```

After you connect to the Raspberry Pi over SSH, run the following commands:

```
iex(1)> VEML6030.start_link()
{:ok, #PID<0.1278.0>}

iex(2)> VEML6030.get_measurement() # Hand covering the sensor
6.4512

iex(3)> VEML6030.get_measurement() # Hand not covering the sensor
71.6544
```

As you can see, we're able to get sensor readings from our VEML6030 GenServer without any issues. The only problem here is that we need to manually start our GenServer process from the IEx session and nothing is in place to supervise it in the event that it crashes. Luckily OTP provides us all the constructs necessary to have our GenServers be fault tolerant and highly available. Let's tackle that in the next section as we complete the configuration and glue code in the sensor_hub firmware project.

Build the Firmware Project

The sensor_hub subproject is the head honcho of the ponchos. Firmware projects exist for each major configuration of a project. For example, if you decide to deploy sensor hubs around your home, you might opt for a wind gauge on a Grisp[2] outside but an air quality sensor on a Raspberry Pi Zero inside. If so, you'd create a subproject for each one.

Each firmware subproject is responsible for collecting dependencies, configuring the target, providing glue code, and managing the life cycle of the device that it's deployed to. That means you will not see many lib/*.ex files, beyond glue code and the supervisor in application.ex. Most of the code you build will be in config.exs, possibly a few tests in test/*_test.exs, and the dependencies in mix.exs.

The sensor_hub project will handle dependencies, manage configuration, provide glue code, and implement the application life cycle. These are the tasks that we'll need to do:

---

2.   https://www.grisp.org

- Add the path and hex dependencies the project needs.

- Provide configuration for the project.

- Create sensor wrappers so each sensor will present a uniform interface in lib/sensor_hub/sensor.ex.

- Start each of the sensor dependencies and the HTTP data publisher.

Let's start with a bit of common configuration that will establish our application life cycle and set the network name of our device.

Configure the Name

In the spirit of configuring our project, let's set the name. Open up config/target.ex in the sensor_hub project. Remember, this is the configuration file that works on all targets except the host. Let's name the node hub to keep it short and descriptive:

```
config :mdns_lite,
  ...comments...
  host: [:hostname, "hub"],
```

From now on, you'll refer to the Nerves device as hub.local rather than nerves.local. After you upload the completed application and the target restarts, you'll use the command mix upload hub.local to upload firmware and ssh hub.local to connect to the device.

With a shiny new name in hand, it's time to configure the first few sensor dependencies.

Include Dependencies for Sensors

In the previous section we added our veml6030 subproject as a path dependency in our sensor_hub subproject. That is what allowed us to query the sensor for light measurements while we were connected to the device over SSH. You may be wondering what we'll do to connect to the other two sensors daisy-chained together (the SGP30 and BME680 sensors). To answer that question we'll turn to the Elixir ecosystem and lean on open source packages published to Hex.

The BME680 sensor already has a couple dependencies on Hex. The best one as of this writing is the bmp280[3] project (this project covers a wide array of Bosch sensors including our BME680). Make sure you get version 0.2.5 or

---

3. https://hex.pm/packages/bmp280

greater because Nerves co-founder Frank Hunleth has done some work to make the dependency work for this book. (Thanks, Frank!).

The SGP30 sensor also has a package available for it on Hex.[4] At the time of this writing, 0.2.0 was the latest version of this library, so we'll be using that.

Let's go ahead and update our mix.exs file in the sensor_hub subproject to leverage these community libraries. Make sure that your deps function looks like so:

```
defp deps do
  [
    ...

    # Dependencies for all targets except :host
    {:veml6030, path: "../veml6030", targets: @all_targets},
    {:sgp30, "~> 0.2.0", targets: @all_targets},
    {:bmp280, "~> 0.2.5", targets: @all_targets},
    ...
  ]
end
```

With that done, run mix deps.get to fetch all of the new project dependencies. Now we can take the project for a test drive. Burn firmware with mix firmware, and upload it with mix upload hub.local. Next, we'll shell in and play around with the new sensors.

Trying the New Sensors

While the project may not be fully configured yet, we're making progress, and it's our first chance to try out the BME680 and SGP30 sensors. After you've connected to the device via ssh hub.local, run the following commands:

```
iex(1)> SGP30.start_link([])
{:ok, #PID<0.1309.0>}

iex(2)> SGP30.state()
%SGP30{
  address: 88,
  co2_eq_ppm: 400,
  ethanol_raw: 16681,
  h2_raw: 12746,
  i2c: #Reference<0.1234438711.268828680.114585>,
  serial: 22277397,
  tvoc_ppb: 0
}

iex(3)> BMP280.start_link([i2c_address: 0x77, name: BMP280])
{:ok, #PID<0.1312.0>}
```

---

4. https://hex.pm/packages/sgp30

```
iex(4)> BMP280.read(BMP280)
{:ok,
 %BMP280.Measurement{
   altitude_m: 70.54530262376304,
   dew_point_c: 8.70336750857366,
   gas_resistance_ohms: 1616.047839376146,
   humidity_rh: 37.11383545983376,
   pressure_pa: 99166.5629969961,
   temperature_c: 24.274734566038205,
   timestamp_ms: 245233
 }}
```

It all works, but each sensor packs up its information differently. The SGP30 sensor returns a struct, the BMP280 returns a result tuple, and our VEML6030 sensor library returns a single value. Let's fix that problem with a sensor wrapper.

Normalize Sensor Measurements with Glue Code

Let's write a few functions to normalize the sensor measurements so we can read each sensor in the same way and convert them to measurements that all have the same format. Each sensor will have a struct with the fields it measures, a read function to measure them, a convert function to normalize those measurements, and a name, which will be the module that implements the sensor. In the sensor_hub firmware project, let's open up a new file at lib/sensor_hub/sensor.ex and start with the struct definition:

```
defmodule SensorHub.Sensor do
  defstruct [:name, :fields, :read, :convert]
end
```

Each sensor will track the fields that it measures and will have functions that know how to read and write to each sensor. Let's now add the constructor for this struct as well as the field reading functions:

```
defmodule SensorHub.Sensor do
  ...

  def new(name) do
    %__MODULE__{
      read: read_fn(name),
      convert: convert_fn(name),
      fields: fields(name),
      name: name
    }
  end

  def fields(SGP30), do: [:co2_eq_ppm, :tvoc_ppb]
  def fields(BMP280), do: [:altitude_m, :pressure_pa, :temperature_c]
  def fields(VEML6030), do: [:light_lumens]
end
```

The constructor creates a new sensor, tracking all of the data we'll later need to interact with it or measure it. The name and fields attributes track the name of the sensors and measurements it collects. We use pattern matching to extract the desired fields, depending on the sensor type. We'll also need to interact with each sensor, so let's write functions that know how to read a measurement, like so:

```elixir
defmodule SensorHub.Sensor do
  ...

  def read_fn(SGP30), do: fn -> SGP30.state() end
  def read_fn(BMP280), do: fn -> BMP280.measure(BMP280) end
  def read_fn(VEML6030), do: fn -> VEML6030.get_measurement() end

  def convert_fn(SGP30) do
    fn reading ->
      Map.take(reading, [:co2_eq_ppm, :tvoc_ppb])
    end
  end

  def convert_fn(BMP280) do
    fn reading ->
      case reading do
        {:ok, measurement} ->
          Map.take(measurement, [:altitude_m, :pressure_pa, :temperature_c])

        _ ->
          %{}
      end
    end
  end

  def convert_fn(VEML6030) do
    fn data -> %{light_lumens: data} end
  end
end
```

Like our fields/1 function, we leverage pattern matching for these functions to calculate the values we need for each of the sensor types. You might recognize these conversion functions because we used the same techniques while we were experimenting with our device via our SSH session. Finally, we can abstract out a measurement with a convenience function, like this:

```elixir
defmodule SensorHub.Sensor do
  ...

  def measure(sensor) do
    sensor.read.()
    |> sensor.convert.()
  end
end
```

That's all our glue code needs to do. We take a sensor, read a value, and then convert the value to a valid measurement. We can use these new functions to simplify our sensor interaction work. Let's test all of this out by creating an updated firmware, uploading it, and SSHing into our device. After you connect to your device, run the following:

```
iex(1)> SGP30.start_link([])
{:ok, #PID<0.1343.0>}

iex(2)> BMP280.start_link([i2c_address: 0x77, name: BMP280])
{:ok, #PID<0.1345.0>}

iex(3)> alias SensorHub.Sensor
SensorHub.Sensor

iex(4)> gas = Sensor.new(SGP30)
%SensorHub.Sensor{
  ...
}

iex(5)> environment = Sensor.new(BMP280)
%SensorHub.Sensor{
  ...
}

iex(6)> Sensor.measure(gas)
%{co2_eq_ppm: 413, tvoc_ppb: 4}

iex(7)> Sensor.measure(environment)
%{
  altitude_m: 71.55274464815629,
  pressure_pa: 99154.70173348083,
  temperature_c: 25.223801961747085
}
```

Everything is working splendidly. Thus far, the firmware subproject contains code for dependencies, configuration, and a tiny amount of glue code. If we wanted, we could build out tiny sensor hubs for custom Raspberry Pis, all with their own set of custom sensors in firmware projects of their own. Now all we need to do is write the code to manage the life cycle of our devices.

Managing the Life Cycle

The application.ex file is often the first module called by Elixir within a new application. It implements a supervisor and is a convenient point for attaching initial startup features. For the purposes of this project, we also need to start the sensors and server publisher services (we'll tackle the publisher in the next chapter). Most of the configuration will go in the application.ex file as arguments to each of the new GenServers we start.

Setting up the Supervision Tree

Open up lib/sensor_hub/application.ex in the sensor_hub subproject, and let's fill out the small functions that will start the pieces of the firmware project. Let's begin with the children/1 function (specifically the version that pattern matches on _target):

```elixir
defmodule SensorHub.Application do
  ...

  def children(_target) do
    # The sensors will fail on the host so let's
    # only start them on the target devices.
    [
      {SGP30, []},
      {BMP280, [i2c_address: 0x77, name: BMP280]},
      {VEML6030, %{}}
    ]
  end

  ...
end
```

This code should look slightly familiar. If you recall from our IEx sessions, when we started the sensor GenServers manually, we had to pass certain options to their start_link/1 functions for them to work properly. Instead of doing that ourselves manually on application start, we'll let our application supervisor take care of that now.

Let's also update our start/2 callback and ensure that it looks like so:

```elixir
defmodule SensorHub.Application do
  ...

  @impl true
  def start(_type, _args) do
    # See https://hexdocs.pm/elixir/Supervisor.html
    # for other strategies and supported options
    opts = [strategy: :one_for_one, name: SensorHub.Supervisor]

    children = children(target())

    Supervisor.start_link(children, opts)
  end

  ...
end
```

As we can see here, our supervisor is provided the list of child processes, depending on what target we're running on. If you look at the Nerves-generated code, the version of children/1 that matches on :host has no child processes, while our Raspberry Pi target contains all of our sensors (this value is provided by the target/0 function). This means that any time we run this project on our workstation, we won't be starting up any of our sensor GenServers, which makes sense given that the sensors are only supposed to run when the project is running on our Raspberry Pi.

With our application life-cycle code set up in our supervisor, it's time to test it all out and make sure that things start up and behave as expected.

Trying It Out

With our code up-to-date, it's time to create an up-to-date firmware via mix firmware, upload the firmware with mix upload hub.local, and then connect to the device with ssh hub.local. After we connect to the device, we can run the following commands to introspect the device and ensure that our sensor GenServers are up and running:

```
iex(1)> Supervisor.which_children(SensorHub.Supervisor)
[
  {VEML6030, #PID<0.1287.0>, :worker, [VEML6030]},
  {BMP280, #PID<0.1286.0>, :worker, [BMP280]},
  {SGP30, #PID<0.1285.0>, :worker, [SGP30]}
]
iex(2)> alias SensorHub.Sensor
SensorHub.Sensor
iex(3)> BMP280 |> Sensor.new() |> Sensor.measure()
%{
  altitude_m: 77.29732484024902,
  pressure_pa: 99087.08904119114,
  temperature_c: 25.63860611162145
}
```

By running the which_children/1 call in IEx, we're able to see what child processes are under the provided supervisor. As you can see, our Nerves application started up all of the GenServers that were specified in the application.ex file, and you were even able to interact with the BMP280 sensor to get sensor data.

Your Turn

In this chapter, we took what we learned from the previous sections and put it all together to create a Nerves application that starts up by itself, initializes all of the sensor hardware, and refreshes sensor measurements automatically.

What You Built

You started the chapter by creating the veml6030 subproject and creating the stateless components to work with your light sensor. These stateless components were derived from the VEML6030 spec sheet and were needed to configure and communicate with the sensor. Once you had these things in place, you added a stateful element to the mix—namely the VEML6030 GenServer module. This GenServer would regularly poll the sensor and store the results in its state. You could then read from this state at any point to get an up-to-date reading on the ambient light in the room.

After creating the veml6030 sensor subproject, you were able to lean on the Elixir and Nerves ecosystem to pull down libraries for working with the additional weather station sensors. You then configured your application supervision tree to start up all of your sensors on device init, and added some glue code to make it easy to fetch data from all of the sensors.

Why It Matters

This chapter walked through exactly how to structure your Nerves applications so that they are bulletproof and production ready. While SSHing into devices and configuring them ad hoc is acceptable for development and experimentation, we need to leverage the OTP available to us to create a reliable and fault-tolerant IoT experience. By using GenServers and Supervisors, we were able to accomplish just that.

What's Next

Now that your Nerves application is almost complete, it's time to set up a Phoenix REST API so that you can publish and persist your sensor data to PostgreSQL+TimescaleDB. Once you have an HTTP server up and running, you'll revisit your Nerves application and add a data publisher subproject to the poncho project, similarly to how you created the veml6030 subproject.

CHAPTER 4

Publishing Sensor Data

With the hardware side of the project almost complete, it's time to create a Phoenix-powered REST server so that you can persist your weather station data. SSHing into the device in an "on-demand" fashion is great for validating that things are working, but in a production-grade application, you'll need a database to house all that data. For this particular project (as well as many other IoT applications), the database of choice is a time-series database. Luckily PostgreSQL has a time-series extension called TimescaleDB, so you can use all of the great Elixir PostgreSQL tooling without any issues (namely Ecto[1]).

The only things that you'll need to have installed on your workstation are Docker[2] and Docker Compose.[3] Once you install both of those for your specific platform, you're ready to dive in!

Setting up Docker Compose

While proficiency with Docker and Docker Compose isn't required, if you're interested in learning more about how Docker works and why it's a useful piece of technology, we suggest going through the Docker documentation[4] and familiarizing yourself with some of the terminology and concepts surrounding containerized applications. In short, you can think of containers as very slim virtual machines (or VMs for short). The key difference is that VMs have separate kernels for each instance, while containers share the underlying operating system kernel. With that being said, let's start off by standing up PostgreSQL so that your Phoenix application has a database to talk to.

---

1. https://hex.pm/packages/ecto
2. https://docs.docker.com/get-docker/
3. https://docs.docker.com/compose/install/
4. https://docs.docker.com/get-started/overview/

Adding PostgreSQL to the Stack

You'll first want to create a vanilla Phoenix application so that you have somewhere to put your docker-compose.yml configuration file. If you don't already have the Phoenix project generator installed on your machine, do so by running mix archive.install hex phx_new 1.5.8. After that is in place, find a directory on your workstation to create your Phoenix application by running the following:

```
$ mix phx.new weather_tracker \
--binary-id --no-webpack --no-html --no-gettext --no-dashboard
```

This command will create a vanilla Phoenix application with binary IDs, no front-end related tooling, no LiveDashboard, and no internationalization tooling. You can enable these disabled items as you see fit, but given that this is strictly a RESTful service, these other components aren't necessary.

With your new vanilla application in place, go ahead and run cd weather_tracker && touch docker-compose.yml to change into the application directory and create the docker-compose.yml file. Using your editor now, add the following to the docker-compose.yml file, and we'll walk through what each bit does:

```yaml
version: '3.3'

services:
  postgres:
    image: timescale/timescaledb:2.1.0-pg13
    ports:
      - '5432:5432'
    volumes:
      - postgres-data:/var/lib/postgresql/data
    environment:
      POSTGRES_PASSWORD: postgres
      POSTGRES_USER: postgres
volumes:
  postgres-data: {}
```

You'll notice that the YAML file has two primary sections: the services and volumes. The services key is used to tell Docker what services it should start as part of the stack, while the volumes key is used to tell Docker what volumes need to be allocated to the running containers. As you can see, the postgres-data entry is referenced in the volumes section of the postgres definition. This enables us to keep our database data across container restarts so that we don't lose the state of our database. You'll also see that we're leveraging the timescale/timescaledb:2.1.0-pg13 image for our postgres service. This PostgreSQL image comes with TimescaleDB pre-installed so that you have all the time-series tooling at your disposal without having to worry about installing or configuring anything yourself. With that in place, you're ready to fire up the Docker Compose stack.

Starting the Docker Compose Stack

All that you have to do now is run docker-compose up from the terminal (assuming you are in the same directory as the docker-compose.yml file), and your PostgreSQL container should start right up and begin outputting logs:

```
$ docker-compose up
...
postgres_1  | PostgreSQL init process complete; ready for start up.
```

And just like that, you have PostgreSQL up and running with time-series data support! Let's get back into Elixir land and work on the Phoenix application. You can leave Docker Compose up and running, given that you'll be connecting to it shortly from your Phoenix application. If you do need to shut it down for whatever reason, all you need to do is press Control+C.

Creating the Phoenix Application

To persist our data in PostgreSQL, we'll need to execute some database migrations so that we have the necessary table in place to store our data. Do that by running the following:

```
$ mix ecto.gen.migration set_up_weather_data_table
```

That command should have generated a file in the priv/repo/migrations directory. Open up that file and set it up so that it looks like this:

weather_tracker/priv/repo/migrations/20210504160714_set_up_weather_data_table.exs
```
defmodule WeatherTracker.Repo.Migrations.SetUpWeatherDataTable do
  use Ecto.Migration

  def up do
    execute("CREATE EXTENSION IF NOT EXISTS timescaledb")

    create table(:weather_conditions, primary_key: false) do
      add :timestamp, :naive_datetime, null: false
      add :altitude_m, :decimal, null: false
      add :pressure_pa, :decimal, null: false
      add :temperature_c, :decimal, null: false
      add :co2_eq_ppm, :decimal, null: false
      add :tvoc_ppb, :decimal, null: false
      add :light_lumens, :decimal, null: false
    end

    execute("SELECT create_hypertable('weather_conditions', 'timestamp')")
  end
```

```
  def down do                        .
    drop table(:weather_conditions)
    execute("DROP EXTENSION IF EXISTS timescaledb")
  end
end
```

This particular migration is split into two parts, the up/0 function and the down/0 function. The up/0 function will migrate the database forward and put in place the items that you define, while the down/0 function will roll the database backward, removing all the items that you define. In this particular case, we want our migration to create a new table called weather_conditions, with a column for each of the sensor data points that we'll be collecting (temperature, altitude, ambient light, and so on). The migration will also enable the timescaledb extension for the application database and will finally convert the weather_conditions table to a time-series table (create_hypertable is a function call provided to us by the TimescaleDB extension to turn regular PostgreSQL tables into time-series capable hypertables).[5]

With the migration in place, all you need to do is run the following:

```
$ mix ecto.setup
The database for WeatherTracker.Repo has been created

12:23:18.621 [info]  == Running 20210504160714 WeatherTracker.Repo.Mig...
12:23:18.622 [info]  execute "CREATE EXTENSION IF NOT EXISTS timescaledb"
12:23:18.623 [info]  extension "timescaledb" already exists, skipping
12:23:18.623 [info]  create table weather_conditions
12:23:18.626 [info]  execute "SELECT create_hyper..."

12:23:18.629 [info]  == Migrated 20210504160714 in 0.0s
```

Let's shift our focus now to our Phoenix context that will be responsible for interacting with this database table.

Creating our Ecto Schema

To handle instances of weather_conditions coming from the database, we'll need to have an Ecto Schema in place to map 1:1 with the fields that you defined in the migration. To that end, create a file in lib/weather_tracker/weather_conditions/ called weather_condition.ex. This file will contains all the Ecto related parts so that we can read and write weather condition data to the database. In that file put the following:

```
defmodule WeatherTracker.WeatherConditions.WeatherCondition do
  use Ecto.Schema
  import Ecto.Changeset
```

---

5.   https://docs.timescale.com/api/latest/hypertable/create_hypertable/

```
@allowed_fields [
  :altitude_m,
  :pressure_pa,
  :temperature_c,
  :co2_eq_ppm,
  :tvoc_ppb,
  :light_lumens
]
@derive {Jason.Encoder, only: @allowed_fields}
@primary_key false
schema "weather_conditions" do
  field :timestamp, :naive_datetime
  field :altitude_m, :decimal
  field :pressure_pa, :decimal
  field :temperature_c, :decimal
  field :co2_eq_ppm, :decimal
  field :tvoc_ppb, :decimal
  field :light_lumens, :decimal
end
end
```

This Ecto Schema module contains fields for all of the sensor data that we have at our disposal on the Nerves device and also adds a timestamp so that we know exactly when the entry was added to the database. You'll also notice that the schema is annotated with @primary_key false. The reason for this is that there's not much of a use case for fetching data out of the database by a unique identifier. The timestamp field is the key index in this table, and it's what you used in your migration to tell TimescaleDB how to generate the hypertable. The @derive attribute is used to instruct Jason (the JSON serialization/deserialization library) how it should encode the %WeatherCondition{} struct into JSON so that it can be conveniently returned to the requesting application (in most cases your response payload will be very lightweight, but it's useful in this application for debugging purposes, as you'll see). You'll also need to add {:decimal, "~> 2.0.0"} to your mix.exs file since we're using the :decimal type for all of the sensor measurement fields.

Next we'll need to add a changeset function to the schema module so that we can cast and validate incoming data in preparation for writing to the database. Append the following to the WeatherCondition schema module:

```
defmodule WeatherTracker.WeatherConditions.WeatherCondition do
  ...
  def create_changeset(weather_condition = %__MODULE__{}, attrs) do
    timestamp =
      NaiveDateTime.utc_now()
      |> NaiveDateTime.truncate(:second)
```

```
    weather_condition
    |> cast(attrs, @allowed_fields)
    |> validate_required(@allowed_fields)
    |> put_change(:timestamp, timestamp)
  end
end
```

And with that, our incoming data can be cast appropriately and validated for required fields, and you'll also have the timestamp created automatically for you when the changeset function is invoked. With that in place, we can move on to creating our Phoenix context module so that data can be inserted into the database.

Creating the Phoenix Context Module

Given that our Phoenix application will only be writing to the PostgreSQL (Grafana will be taking care of our presentation layer requirement in the next chapter), our Phoenix context module will only have a single function in it. That function will leverage the WeatherCondition schema that you created in the previous section and insert it into the database via your Repo module. To create the the Phoenix WeatherCondition context module, create a file at lib/weather_tracker/weather_conditions.ex with the following contents:

weather_tracker/lib/weather_tracker/weather_conditions.ex
```
defmodule WeatherTracker.WeatherConditions do
  alias WeatherTracker.{
    Repo,
    WeatherConditions.WeatherCondition
  }

  def create_entry(attrs) do
    %WeatherCondition{}
    |> WeatherCondition.create_changeset(attrs)
    |> Repo.insert()
  end
end
```

If the data passed in via attrs is invalid, then this function will return an error tagged tuple in the shape of {:error, %Ecto.Changeset{}}. If all goes well and the data is inserted into the database, you'll get a {:ok, %WeatherCondition{}} tagged tuple back.

With the Phoenix context ready to go, all that's left for the server-side application is to create a controller to handle incoming API requests and to then hook up that controller in our router.

Setting up the Phoenix API

Similarly to our Phoenix context, our WeatherConditionsController will be very lightweight and will be focused on creating entries in the database, so we'll only have a POST handler (as per the REST conventions where POST is used to create data in a service). One thing to note is that we won't be performing any kind of input validation at the controller layer, given that we validate the input payload at the Ecto Schema layer.

Another item that should be called out here is that our API is not performing any kind of validation to check that the incoming data is coming from a trusted source. Given that we're running both the Nerves device and the Phoenix API on the same LAN, this problem of request authenticity isn't something that we'll tackle in this book. If you plan on deploying this project and having it be accessible over the public internet, you'll want to have something in place in the controller to validate that the incoming requests are coming from trusted sources. For that purpose you may want to investigate using an HMAC and signing your outgoing Nerves payloads with a shared secret key that the server can also validate.[6]

With that small disclaimer out of the way, let's get the controller in place. Create a file weather_conditions_controller.ex in lib/weather_tracker_web/controllers, with the following contents:

weather_tracker/lib/weather_tracker_web/controllers/weather_conditions_controller.ex
```elixir
defmodule WeatherTrackerWeb.WeatherConditionsController do
  use WeatherTrackerWeb, :controller

  require Logger

  alias WeatherTracker.{
    WeatherConditions,
    WeatherConditions.WeatherCondition
  }

  def create(conn, params) do
    IO.inspect(params)

    case WeatherConditions.create_entry(params) do
      {:ok, weather_condition = %WeatherCondition{}} ->
        Logger.debug("Successfully created a weather condition entry")

        conn
        |> put_status(:created)
        |> json(weather_condition)
```

---

6. https://dashbit.co/blog/how-we-verify-webhooks

```
    error ->
      Logger.warn("Failed to create a weather entry: #{inspect(error)}")

      conn
      |> put_status(:unprocessable_entity)
      |> json(%{message: "Poorly formatted payload"})
    end
  end
end
```

Our controller is leveraging the standard Phoenix.Controller[7] function calls as well as the context module that we wrote in the previous section. When the context module successfully creates a weather condition entry in the database, we return a 201 to the client. If you fail due to malformed data, we log out the errors and return a 422 back to the client. This will be helpful when you implement your data publisher in the Nerves application so that you can debug any API issues.

With the controller in place, the last thing that needs to be done is to add an entry to our router.ex file so our controller will be invoked when the route is invoked. Open up lib/weather_tracker_web/router.ex and ensure that it looks like so:

weather_tracker/lib/weather_tracker_web/router.ex
```
defmodule WeatherTrackerWeb.Router do
  use WeatherTrackerWeb, :router

  pipeline :api do
    plug :accepts, ["json"]
  end

  scope "/api", WeatherTrackerWeb do
    pipe_through :api

    post "/weather-conditions", WeatherConditionsController, :create
  end
end
```

Since the Phoenix application is only serving RESTful API calls, you'll only have a single :api pipeline that validates that the incoming payload is of type application/json. Further down you'll notice that WeatherConditionsController is only invoked whenever a POST request is made to /api/weather-conditions. With that in place, you can run mix phx.server and start up the Phoenix server! Leave the server running, as you'll be publishing metrics soon, and make sure that you know the IP address of your workstation so you can send your metrics to the correct server.

---

7. https://hexdocs.pm/phoenix/Phoenix.Controller.html

All that needs to be done now is to configure the Nerves application to publish metrics to the Phoenix server and we're ready to view all of our time-series data. Let's dive back into the Nerves app and get that up and running.

Publishing Metrics

With our Phoenix server now listening for incoming requests, we'll want to have something running on the Raspberry Pi that regularly collects and publishes metrics from the sensors. Similarly to how we split out the VEML6030 sensor code into a separate poncho subproject, you'll also want to split out the publisher code into its own subproject. If you navigate back to the sensor_hub_poncho project directory and run mix new publisher, you'll create a new vanilla Mix project. With that in place, let's start setting up the publisher code.

Creating the Data Publisher

After navigating to the publisher directory, open up the lib/publisher.ex file. This subproject will be laser focused on only publishing sensor data, and so this file will be the only one you'll be editing. Much like how the VEML6030 module was a GenServer that was started in the firmware subproject, our Publisher module will follow a similar pattern. At a high level, the Publisher module will also be a GenServer that regularly publishes sensor data to a configured host. Let's start off by implementing the start_link/1 and init/1 functions:

```elixir
defmodule Publisher do
  use GenServer

  require Logger

  def start_link(options \\ %{}) do
    GenServer.start_link(__MODULE__, options, name: __MODULE__)
  end

  @impl true
  def init(options) do
    state = %{
      interval: options[:interval] || 10_000,
      weather_tracker_url: options[:weather_tracker_url],
      sensors: options[:sensors],
      measurements: :no_measurements
    }

    schedule_next_publish(state.interval)

    {:ok, state}
  end
```

```
  defp schedule_next_publish(interval) do
    Process.send_after(self(), :publish_data, interval)
  end
end
```

The start_link/1 function is used to start the Publisher GenServer and give it a unique name so that this singleton process can be referenced by the module name. In the init/1 function, you'll see that you are extracting certain fields from the provided options and building the state for the GenServer. The call to the schedule_next_publish/0 function schedules a message that will be sent to self() (the GenServer PID) ten seconds into the future (or however long you decide to configure the interval). We'll get into why this is important shortly.

Alex says:
Why Not Publish Sensor Data Every Second?

If you recall from Chapter 3, you made a call in your VEML6030 GenServer init/1 callback to refresh the sensor data every second via: :timer.send_interval(1_000, :measure). You may be wondering why the Publisher module only sends data to the server every ten seconds when we have measurements captured at a one-second resolution?

The reason for this is that for this particular project, a ten-second resolution is more than sufficient. Weather conditions are relatively slow-moving phenomena and over sampling the sensor would result in excess data being stored in the database. For example, if you were to store a collection of measurements every second, that would result in about 86,400 rows in the database every single day that the Nerves device is capturing and publishing data. For some use cases this may be required, but for this application it's not necessary.

You can experiment with this on your own by adjusting the data publish interval to see what your data looks like in the database. Or you can even sample the sensor every second for ten seconds, average the results over that time window to smooth out the sensor readings over time, and publish only a single data point for that ten-second time window. We encourage you to experiment with the data collection and data massaging aspects to see how the end result is altered.

What needs to be taken care of now is the handle_info/2 callback to service the :publish_data message. Let's add that next along with its supporting private functions:

```
defmodule Publisher do
  ...

  @impl true
  def handle_info(:publish_data, state) do
    {:noreply, state |> measure() |> publish()}
  end
```

```
defp measure(state) do
  measurements =
    Enum.reduce(state.sensors, %{}, fn sensor, acc ->
      sensor_data = sensor.read.() |> sensor.convert.()
      Map.merge(acc, sensor_data)
    end)

  %{state | measurements: measurements}
end

defp publish(state) do
  result =
    :post
    |> Finch.build(
      state.weather_tracker_url,
      [{"Content-Type", "application/json"}],
      Jason.encode!(state.measurements)
    )
    |> Finch.request(WeatherTrackerClient)

  Logger.debug("Server response: #{inspect(result)}")

  schedule_next_publish(state.interval)

  state
end
end
```

The handle_info/2 callback is short and delegates most of the work to the supporting private functions. The private functions return an updated state map so you can compose them with the pipe operator. The measure/1 function is responsible for aggregating metrics from all of the configured sensors and then updating the GenServer state with the captured measurements. It then passes that updated state to the next private function publish/1, which leverages the HTTP library Finch to send the data to the configured URL (the data is converted to JSON using the Jason library).

One line to that you should take note of is the last element of the pipe chain Finch.request(WeatherTrackerClient). This function call is what makes the HTTP request to our server. To make this HTTP call, Finch requires a running pool of connections (in that case, the connection pool is addressable by the name WeatherTrackerClient). To use the Finch HTTP library, you'll need to update the mix.exs file in the publisher subproject and include both {:finch, "~> 0.6.3"} and {:jason, "~> 1.2.2"} in the dependency list. Let's now add the data publishing GenServer and the Finch connection pool to the firmware project supervision tree to get this working end-to-end.

Hooking It into the Firmware Project

To wrap up the Nerves side of this application, open up lib/sensor_hub/application.ex in the sensor_hub subproject and update the module to look like so:

```elixir
defmodule SensorHub.Application do
  ...

  alias SensorHub.Sensor

  ...

  def children(_target) do
    [
      {SGP30, []},
      {BMP280, [i2c_address: 0x77, name: BMP280]},
      {VEML6030, %{}},
      {Finch, name: WeatherTrackerClient},
      {
        Publisher,
        %{
          sensors: sensors(),
          weather_tracker_url: weather_tracker_url()
        }
      }
    ]
  end

  defp sensors do
    [Sensor.new(BMP280), Sensor.new(VEML6030), Sensor.new(SGP30)]
  end

  defp weather_tracker_url do
    Application.get_env(:sensor_hub, :weather_tracker_url)
  end

  ...
end
```

As you can see from the updated children/1 function, we now have an entry for the Publisher GenServer module that we wrote in the previous section along with the Finch process that will handle the creation of HTTP connection pools. The last thing that needs to be done now is to update the sensor_hub/config/target.exs file and include a config entry for :weather_tracker_url that you see referenced at the bottom of the SensorHub.Application module.

With that being said, add the following to sensor_hub/config/target.exs and replace <SERVER_IP_ADDRESS> with the IP address of your Phoenix server on the LAN:

```elixir
config :sensor_hub, :weather_tracker_url,
  "http://<SERVER_IP_ADDRESS>:4000/api/weather-conditions"
```

Lastly, you'll need to update the mix.exs file found in the sensor_hub subproject to include the new poncho project dependency. As you've done with the other poncho project dependencies, include the :publisher subproject using the :path option:

```
defp deps do
  [
    # Dependencies for all targets
    ...

    # Dependencies for all targets except :host
    {:publisher, path: "../publisher", targets: @all_targets},
    ...

    # Dependencies for specific targets
    ...
  ]
end
```

All that's left now is to burn a firmware and upload it to the Raspberry Pi. As before, be sure to have the MIX_TARGET environment variable exported, and run the following:

```
$ mix firmware
$ mix upload
$ ssh hub.local
```

After you connect to the device over SSH, run the following in your IEx session to attach to the default Nerves logger back end and see the server responses coming back to your Nerves device:

```
iex(1)> RingLogger.attach()
:ok

19:28:48.411 [debug] Server response: {:ok, %Finch.Response{..., status: 201}}

19:28:58.855 [debug] Server response: {:ok, %Finch.Response{..., status: 201}}

iex(2)> exit()
```

By running RingLogger.attach() on the Nerves device, you can see the debug log messages coming out of the Publisher GenServer to let you know that your device is communicating with the Phoenix application. If you see anything aside from a status: 201, you know that there's an issue with the server-side of the application and you can start debugging from there. If you see something along the lines of the following, then you know that you have an issue with the Nerves application configuration, as it is unable to communicate with the Phoenix server.

```
iex(1)> RingLogger.attach()
:ok
19:33:11.235 [debug] Server response: {:error, %Mint.TransportError{...}}
```

If you do see this, make sure that the IP address that you set in the configuration is the actual IP address of the server on the LAN and that the port number is also set correctly.

Your Turn

In this chapter, we built the Phoenix server application that's used to ingest all of the Nerves sensor data. The sensor data is then stored into Postgres as time-series data, thanks to TimescaleDB.

What You Built

You started off the chapter by creating a Docker Compose stack to start a Postgres container with the TimescaleDB extension pre-installed. You also made sure to have a volume entry in the YAML manifest file so that your time-series data is persisted across application restarts. You then leveraged this Postgres instance through a Phoenix back-end application that you wrote. Once the Ecto migration, schema, Phoenix context, and controller were put together, you were ready to ingest time-series data.

Next, you shifted gears back to the Nerves application and added an additional poncho subproject that was used to publish sensor data at a regular interval. Once you added some minor configuration for the data-publishing GenServer you were able to see from the IEx session logs were flowing back successfully.

Why It Matters

This chapter marks a big milestone since you finally have a technology stack that completely works end-to-end. The Nerves application on the Raspberry Pi is able to capture sensor data and publish it to the Phoenix server over the wireless LAN, and the Phoenix application is able to persist all the data to the database. The amazing part is that you were able to leverage the same programming language across both platforms without having to lean on any escape hatches. Elixir is just as capable on the back end as it is on an embedded device (thanks to Nerves).

What's Next

Now that you have data flowing from the Raspberry Pi to the Phoenix back end, all that's left is to visualize all that data. You probably noticed that there were no read operations in the WeatherConditions Phoenix context. The reason for this is that you'll leverage the amazingly capable visualization service called Grafana. Luckily Grafana runs great inside of a Docker container, so you'll update your Docker Compose stack to include Grafana and effortlessly visualize all of the weather data that your Nerves sensor hub is capturing. With that said, be sure to leave your Nerves sensor hub plugged in somewhere interesting for a few hours so that you'll have ample time-series data to surface through Grafana.

CHAPTER 5

Pulling It All Together

With the Nerves device regularly sending data to our Phoenix application, all that's left is to present all of this time-series data in a dashboard. Luckily, Grafana is both open source and very capable at surfacing data (and in particular time-series data) in configurable visuals like line graphs, pie charts, heatmaps, and gauges, to name a few.

Let's start off by adding Grafana to the Docker compose stack and getting it to fetch data out of the Postgres instance.

Adding Grafana to Docker Compose

Like TimescaleDB, we can easily run Grafana inside of a container and simply mount the appropriate volumes where necessary so that Grafana can persist the dashboard changes across restarts. We'll leverage the Docker image from Docker Hub and add this additional service to our YAML docker-compose.yml manifest file.[1]

Open up docker-compose.yml and add the following:

```
version: '3.3'
services:
  postgres:
    ...
  grafana:
    image: grafana/grafana:8.0.5
    depends_on:
      - postgres
    ports:
      - '3000:3000'
```

---

1. https://hub.docker.com/r/grafana/grafana

```
    volumes:
      - grafana-data:/var/lib/grafana
volumes:
  ...
  grafana-data: {}
```

Similarly to how we defined the postgres service, we define the new grafana service and specify what Docker image should be leveraged for the service. At the time of this writing, version 8.0.5 was the latest version of Grafana. We also specify a depends_on clause to let Docker Compose know this service should be started after Postgres, given the dashboard will depend on reading data out of Postgres. We also specify the service should be accessible on port 3000 of the host workstation, and lastly we have the volume definition in place so we don't lose our dashboards when the Docker Compose stack is restarted.

With that in place, all that's left is to start up the updated Docker Compose stack with docker-compose up. If your Docker Compose stack is still running and collecting data, press Control+C to stop it and then run docker-compose up. After Docker has finished downloading the new container image, navigate to localhost:3000 and we'll start exploring the data (by default the username and password to Grafana are both admin).

Exploring the Data with SQL

After you've navigated to http://localhost:3000, hover on the Configuration button in the left-hand side navigation bar and select Data Sources (you can also navigate directly to http://localhost:3000/datasources as opposed to going through the side menus).

This part of Grafana is used to configure what persistent data sources Grafana can communicate with to retrieve and display data. While there are many supported data sources in Grafana, the one that we're interested in is Postgres. Click the Add data source button and then search for Postgres, as seen in the following image:

After you find it, select it and configure the data source similarly to how the following image shows:

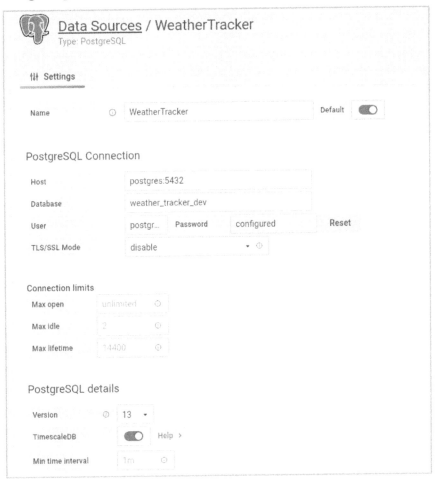

Be sure to toggle the TimescaleDB option to enabled so that Grafana knows that the Postgres instance it is working with supports TimescaleDB time-series queries, and use the same user and password that you did for the Phoenix back end (both the username and password should be postgres if you didn't provide your own credentials). Also be sure to disable TLS/SSL Mode, since this application stack isn't operating over the public internet, and communicating over insecure transports on the LAN is acceptable for the purposes of this project. After all that is done, click the Save & Test button at the bottom and ensure that Grafana reports back Database Connection OK. With the Postgres data source set up, we're ready to explore our Postgres database.

Inspecting Environment CO2

While Grafana can be used to create awesome dashboards in very little time, it also offers a great way to do ad-hoc queries and explore the data present in your configured data sources. Let's give this a test-drive by hovering on the compass icon on the side navigation and clicking the Explore menu item (or go there directly by navigating to http://localhost:3000/explore). Once at the Explore page, fill out the query builder as follows:

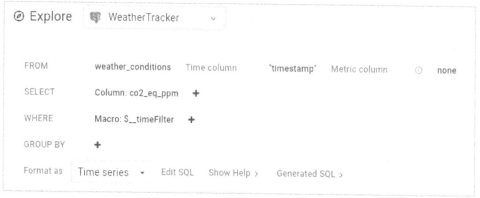

Before previewing the results, let's walk through what we've done with the query builder here. Firstly, we told Grafana to query the weather_conditions table and to specifically SELECT the co2_eq_ppm column from the table. The Time column and WHERE options were left as is, given that the Grafana defaults were sufficient for the query.

With the query built, click the Run query button in the upper-right side of the screen and observe the results. Here are the results that I got after running the query and plotting my captured data:

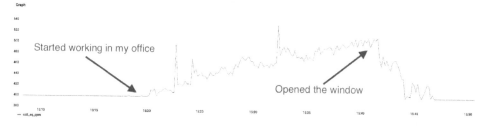

In my particular instance, I had my sensor hub in my office while writing this chapter and was observing the CO2 levels that the SGP30 sensor was measuring. It was interesting to see that while I was in my office without the window open, the CO2 levels were increasing at a regular rate. As soon as I opened the office window, it only took about five minutes for the CO2 PPM measurement to drop by almost 100 points. Feel free to explore and investigate the other sensor data points and try to correlate the data with events occurring in your environment.

Now that we have a sense for how to explore the time-series data, let's switch over from making ad-hoc database queries to having a dashboard present all of the data points of interest to us automatically.

Creating a Weather Dashboard

Out of the box, Grafana can generate a wide array of visualizations appropriate for different types of data. You can create bar charts, stat panels, gauges, heatmaps, line graphs, and many more. To get comfortable with a few of the visualization types that Grafana offers, you'll be creating a dashboard that leverages the stat panel, gauge, and time-series chart types. The end product for what we'll be building in this chapter will look something like this:

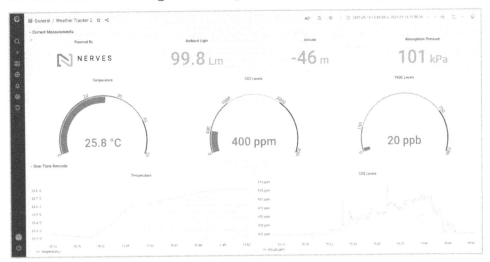

Feel free to deviate from the instructions if you want to make the dashboard your own and want a different look and feel. With that said, let's start off by creating the Current Measurements row of panels.

Adding a Stat Panel

To create your first Grafana dashboard, start off by hovering on the plus icon in the side navigation and selecting Dashboard. After you click that, you'll be presented with a blank dashboard and an option to Add an empty panel or Add a new row. Select Add a new row and then hover on the row and click the gear icon to give it the name of Current Measurements. With that in place, click the button in the upper right labeled Add panel to create a new panel, as shown in the image on page 68.

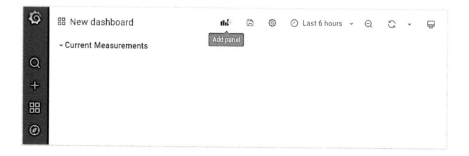

After clicking the Add panel button, you'll be presented with the same options that you had when you first created the dashboard. This time around, click the Add an empty panel button. You should now be at the Edit Panel page so that you can configure your new panel. Given that this will be a state panel, click the drop-down menu in the upper right to change the visualization type (the default selected type is Time Series) to Stat. The stat panel will display the last non-null value from the database for the selected time range. The first thing to do here on the Edit Panel page is to fill out the query builder so that it looks like so:

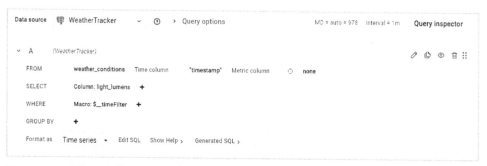

Once you fill out the query builder, you should see your data populate the stat panel above. Feel free to go through the menu on the right and customize the panel, now that it's displaying data. Be sure to provide a title for the panel, set the units to Lumens (Lm), and remove the default color threshold if you don't want the stat panel to change colors when the light sensor is picking up a lot of light.

With your stat panel configured, click the Apply button in the upper right and resize the panel by dragging it from the bottom-right corner of the panel. Now that one stat panel has been created, you can easily duplicate it and slightly tweak the query builder to display a different data point. To duplicate the stat panel, click the title for the panel, go to the More... menu, and select Duplicate. You can then edit the duplicated panel by again clicking the title and selecting Edit. In the query builder you can then input altitude_m, for example, to display the altitude (be sure to change the units and the title so you don't confuse yourself as to which panel is which). For reference, the final dashboard visual at the beginning of the section has stat panels for ambient light, altitude, and atmospheric pressure.

With that done, let's play around with the gauge visualization type.

Adding a Gauge

Similarly to before, start off by clicking the Add panel button in the upper left, and move the blank panel under the stat panels that you just created. Then click Add an empty panel and select Gauge from the visuals drop-down menu. Similarly to the stat panel, fill out the query builder, but this time select the temperature_c column.

After filling out the title and units fields, go down to the Thresholds section and fill it out like so:

You can set the color for each threshold by clicking the colored circle and selecting a color for that threshold. Also be sure to set the Min to 0, Max to 50, and Show threshold labels to enabled. With that done, you should have a gauge that looks something like this:

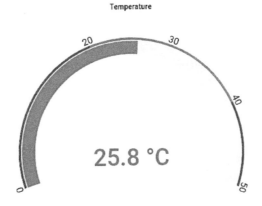

As with the stat panels, you can duplicate the gauge that you just created, and edit the selected column to display gauges for CO2 and TVOC measurements. Feel free to reference the screenshot at the beginning of the section to set the thresholds for these panels. All that's left now to complete the weather station dashboard is to plot some measurements over time. Let's do that next.

Adding a Line Graph

As before, start out by clicking the Add panel button, and select Add a new row. Give the row a title of Over Time Records, and drag it down to the bottom of the dashboard (you'll have to collapse the row prior to dragging it or else it will stay fixed in place). With that done, press the Add panel button again and then Add an empty panel. Given that Time series is the default visualization type in Grafana, you won't have to change anything to plot time-series data in a line graph.

As with the previous visualizations, fill out the query builder and input temperature_c as the desired column to plot. When that's done, enter a title for the panel and be sure to set the unit to Celsius (°C). Once that's done, you should have a visualization that looks something like so:

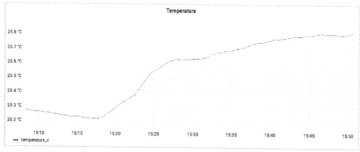

Once you have the time-series graph set up how you like, you can go ahead and duplicate it to plot the CO2 and TVOC levels. While the stat and gauge visuals only show the last non-null value, the time-series graphs will plot all the persisted data points within the selected time range. You can customize the visible time window by selecting the desired range in the drop-down menu in the upper-right corner.

Wrapping Up

From the dashboard page you can save the dashboard's current state by clicking the Save dashboard button next to the Add panel button, or you can add additional visualization panels. From here, it comes down to what data points are of particular interest to you and how you want to surface that information.

You can even go so far as to have Grafana send notifications any time environmental readings hit a certain threshold.[2]

If you want to import the dashboard that was presented at the beginning of the section, be sure to check out the GitHub source for this project.[3] The JSON dashboard definition in that repository can be imported using the Grafana import tool that can be found at http://localhost:3000/dashboard/import. You can also export your dashboard by clicking the gear icon next to the time range drop-down menu. Once in the Dashboard settings page, you can go to the JSON Model tab and extract the JSON definition of the dashboard. This can then be saved into source control so that you can load up this dashboard into any future Grafana deployments.

Your Turn

Congratulations for setting up your Nerves-powered weather station and getting it all running end-to-end! Now that you have a complete weather station solution in place (from data collection to presentation), it's easy to see why Elixir and Nerves are such a compelling technology stack for IoT applications and how well they work for this class of problems. From Elixir's GenServer and supervision constructs to Nerve's out-of-the-box ability to upload device firmware images wirelessly, the whole development experience has been tuned for optimal productivity and reliability.

What You Built

In this chapter, you wrapped up your weather station project by adding Grafana to the technology stack. With Grafana in place and connected to Postgres+TimescaleDB, you were able to create dashboards that could surface the time-series measurements captured by the Nerves device. With this last piece of the puzzle in place, you have an end-to-end solution that can capture, persist, and view weather data collected by your Raspberry Pi.

Why It Matters

This is an important milestone in the project as it showcases the capabilities of the Nerves-powered weather station as a complete IoT solution. Using the Elixir programming language, you were able to capture and publish metrics from an embedded device. And using the same programming language, you were able to persist that data into a time-series database. This is a testament

---

2.   https://grafana.com/docs/grafana/latest/alerting/notifications/
3.   https://github.com/akoutmos/nerves_weather_station

to how well suited the Elixir programming language and Erlang virtual machine are to a wide array of problem domains. The fundamental promises of reliability, fault tolerance, and a productive developer experience are on display in the context of an embedded device and of a back-end API.

What's Next

Now that you have a productive development workflow and a working Nerves device, where you go next is up to you. You can hook up other I2C Qwiic sensors or LCD screens,[4] you can deploy a fleet of Nerves devices and aggregate metrics across all the devices, or you can even try and deploy the same Nerves application to other embedded devices like the BeagleBone Black.[5] With an understanding of how to develop, organize, and deploy Nerves applications, you should be comfortable with a wide array of embedded Elixir+Nerves projects. We encourage you to explore the many facets of the Nerves framework and build more robust and reliable Nerves applications.

---

4. https://www.sparkfun.com/qwiic
5. https://beagleboard.org/black

Thank you!

We hope you enjoyed this book and that you're already thinking about what you want to learn next. To help make that decision easier, we're offering you this gift.

Head on over to https://pragprog.com right now, and use the coupon code BUYANOTHER2022 to save 30% on your next ebook. Offer is void where prohibited or restricted. This offer does not apply to any edition of the *The Pragmatic Programmer* ebook.

And if you'd like to share your own expertise with the world, why not propose a writing idea to us? After all, many of our best authors started off as our readers, just like you. With a 50% royalty, world-class editorial services, and a name you trust, there's nothing to lose. Visit https://pragprog.com/become-an-author/ today to learn more and to get started.

We thank you for your continued support, and we hope to hear from you again soon!

The Pragmatic Bookshelf

Pragmatic Bookshelf

SAVE 30%!
Use coupon code
BUYANOTHER2022

Powerful Command-Line Applications in Go

Write your own fast, reliable, and cross-platform command-line tools with the Go programming language. Go might be the fastest—and perhaps the most fun—way to automate tasks, analyze data, parse logs, talk to network services, or address other systems requirements. Create all kinds of command-line tools that work with files, connect to services, and manage external processes, all while using tests and benchmarks to ensure your programs are fast and correct.

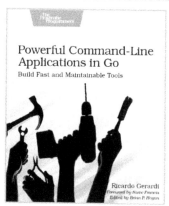

Ricardo Gerardi

(508 pages) ISBN: 9781680506969. $45.95

https://pragprog.com/book/rggo

Pythonic Programming

Make your good Python code even better by following proven and effective pythonic programming tips. Avoid logical errors that usually go undetected by Python linters and code formatters, such as frequent data look-ups in long lists, improper use of local and global variables, and mishandled user input. Discover rare language features, like rational numbers, set comprehensions, counters, and pickling, that may boost your productivity. Discover how to apply general programming patterns, including caching, in your Python code. Become a better-than-average Python programmer, and develop self-documented, maintainable, easy-to-understand programs that are fast to run and hard to break.

Dmitry Zinoviev

(150 pages) ISBN: 9781680508611. $26.95

https://pragprog.com/book/dzpythonic

Concurrent Data Processing in Elixir

Learn different ways of writing concurrent code in Elixir and increase your application's performance, without sacrificing scalability or fault-tolerance. Most projects benefit from running background tasks and processing data concurrently, but the world of OTP and various libraries can be challenging. Which Supervisor and what strategy to use? What about GenServer? Maybe you need back-pressure, but is GenStage, Flow, or Broadway a better choice? You will learn everything you need to know to answer these questions, start building highly concurrent applications in no time, and write code that's not only fast, but also resilient to errors and easy to scale.

Svilen Gospodinov
(174 pages) ISBN: 9781680508192. $39.95
https://pragprog.com/book/sgdpelixir

Testing Elixir

Elixir offers new paradigms, and challenges you to test in unconventional ways. Start with ExUnit: almost everything you need to write tests covering all levels of detail, from unit to integration, but only if you know how to use it to the fullest—we'll show you how. Explore testing Elixir-specific challenges such as OTP-based modules, asynchronous code, Ecto-based applications, and Phoenix applications. Explore new tools like Mox for mocks and StreamData for property-based testing. Armed with this knowledge, you can create test suites that add value to your production cycle and guard you from regressions.

Andrea Leopardi and Jeffrey Matthias
(262 pages) ISBN: 9781680507829. $45.95
https://pragprog.com/book/lmelixir

Help Your Boss Help You

Develop more productive habits in dealing with your manager. As a professional in the business world, you care about doing your job the right way. The quality of your work matters to you, both as a professional and as a person. The company you work for cares about making money and your boss is evaluated on that basis. Sometimes those goals overlap, but the different priorities mean conflict is inevitable. Take concrete steps to build a relationship with your manager that helps both sides succeed.

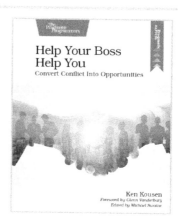

Ken Kousen
(160 pages) ISBN: 9781680508222. $26.95
https://pragprog.com/book/kkmanage

Web Development with Clojure, Third Edition

Today, developers are increasingly adopting Clojure as a web-development platform. See for yourself what makes Clojure so desirable as you create a series of web apps of growing complexity, exploring the full process of web development using a modern functional language. This fully updated third edition reveals the changes in the rapidly evolving Clojure ecosystem and provides a practical, complete walkthrough of the Clojure web stack.

Dmitri Sotnikov and Scot Brown
(468 pages) ISBN: 9781680506822. $47.95
https://pragprog.com/book/dswdcloj3

Hands-on Rust

Rust is an exciting new programming language combining the power of C with memory safety, fearless concurrency, and productivity boosters—and what better way to learn than by making games. Each chapter in this book presents hands-on, practical projects ranging from "Hello, World" to building a full dungeon crawler game. With this book, you'll learn game development skills applicable to other engines, including Unity and Unreal.

Herbert Wolverson
(342 pages) ISBN: 9781680508161. $47.95
https://pragprog.com/book/hwrust

Modern Front-End Development for Rails

Improve the user experience for your Rails app with rich, engaging client-side interactions. Learn to use the Rails 6 tools and simplify the complex JavaScript ecosystem. It's easier than ever to build user interactions with Hotwire, Turbo, Stimulus, and Webpacker. You can add great front-end flair without much extra complication. Use React to build a more complex set of client-side features. Structure your code for different levels of client-side needs with these powerful options. Add to your toolkit today!

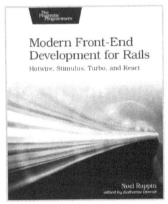

Noel Rappin
(396 pages) ISBN: 9781680507218. $45.95
https://pragprog.com/book/nrclient

The Pragmatic Bookshelf

The Pragmatic Bookshelf features books written by professional developers for professional developers. The titles continue the well-known Pragmatic Programmer style and continue to garner awards and rave reviews. As development gets more and more difficult, the Pragmatic Programmers will be there with more titles and products to help you stay on top of your game.

Visit Us Online

This Book's Home Page
https://pragprog.com/book/passweather
Source code from this book, errata, and other resources. Come give us feedback, too!

Keep Up to Date
https://pragprog.com
Join our announcement mailing list (low volume) or follow us on twitter @pragprog for new titles, sales, coupons, hot tips, and more.

New and Noteworthy
https://pragprog.com/news
Check out the latest pragmatic developments, new titles and other offerings.

Save on the ebook

Save on the ebook versions of this title. Owning the paper version of this book entitles you to purchase the electronic versions at a terrific discount.

PDFs are great for carrying around on your laptop—they are hyperlinked, have color, and are fully searchable. Most titles are also available for the iPhone and iPod touch, Amazon Kindle, and other popular e-book readers.

Send a copy of your receipt to support@pragprog.com and we'll provide you with a discount coupon.

Contact Us

Online Orders:	*https://pragprog.com/catalog*
Customer Service:	*support@pragprog.com*
International Rights:	*translations@pragprog.com*
Academic Use:	*academic@pragprog.com*
Write for Us:	*http://write-for-us.pragprog.com*
Or Call:	+1 800-699-7764

Lightning Source UK Ltd.
Milton Keynes UK
UKHW031815040222
398231UK00006B/14